# Macromolecular Symposia



**163**

WILEY-VCH

pp. 1–250

January 2001

**Macromolecular Symposia** publishes lectures given at international symposia and is issued irregularly, with normally 14 volumes published per year. For each symposium volume, an Editor is appointed. The articles are peer-reviewed. The journal is produced by photo-offset lithography directly from the authors' typescripts.
Further information for authors can be obtained from:
Editorial office "Macromolecular Symposia"
Wiley-VCH
P.O. Box 10 11 61, D-69451 Weinheim,
Germany
or for parcel and courier services: Pappelallee 3, D-69469 Weinheim,
Germany
Tel. +49 (0) 62 01/6 06-2 34 or -2 38; Fax +49 (0) 62 01/6 06-3 09 or 5 10; macromol@wiley-vch.de
http://www.wiley-vch.de/home/macrosymp
Suggestions or proposals for conferences or symposia to be covered in this series should also be sent to the Editorial office at the address above.

**Macromolecular Symposia:**
**Annual subscription rates 2001**
Germany, Austria € 1098 (DM 2147,50); Switzerland SFr 1808; other Europe € 1098; outside Europe US $ 1328.
Macromolecular Package, including Macromolecular Chemistry & Physics (18 issues), Macromolecular Rapid Communications (18 issues), Macromolecular Theory & Simulations (9 issues) is also available. Details on request.
Packages including Macromolecular Symposia and Macromolecular Materials & Engineering are also available. Details on request.
Single issues and back copies are available. Please, inquire for prices.

**Orders** may be placed through your bookseller or directly at the publishers: WILEY-VCH Verlag GmbH, P.O. Box 10 11 61, D-69451 Weinheim, Germany, Tel.: (0 62 01) 6 06–0, Telefax (0 62 01) 60 61 17, Telex 46 55 16 vchwh d. E-mail: subservice@wiley-vch.de

Macromolecular Symposia (ISSN 1022-1360) is published with 14 volumes per year by WILEY-VCH Verlag GmbH, P.O. Box 10 11 61, D-69451 Weinheim, Germany. Air freight and mailing in the USA by Publications Expediting Inc., 200 Meacham Ave., Elmont, NY 11003. Periodicals postage pending at Jamaica, NY 11431. POSTMASTER: send address changes to Macromolecular Symposia, Publications Expediting Inc., 200 Meacham Ave., Elmont, NY 11003.


Printing: Strauss Offsetdruck, Mörlenbach. Binding: Wilh. Osswald, Neustadt

**Symposium**
**of the Division of Macromolecular Chemistry**
**of the GDCh (Gesellschaft Deutscher Chemiker)**

**on**

# Tailormade Polymers

*Merseburg, Germany*
*March 20–21, 2000*

*Symposium Editors*


| | |
|---|---|
| **H. Cherdron** | Wiesbaden |
| **B. Sandner** | Halle |
| **H.U. Schenk** | Ludwigshafen |
| **B. Voit** | Dresden |


*Organizing Committee*


| | |
|---|---|
| **L. Kießling** | Frankfurt/Main |
| **G. Hoffmann** | Ludwigshafen |



ISBN 3-527-30 325-1

The **tables of contents** of the published issues are displayed on the www.

This service as well as further information on our journals can be found at the following www address:

http://www.wiley-vch.de/home/macrosymp

The service is free for everybody.

# Contents of Macromol. Symp. 163

**Tailormade Polymers**
**Merseburg, Germany, 2000**

* The asterisk indicates the name of the author to whom inquiries should be addressed

## Author Index

# Introductory Remarks

to the GDCh Division Macromolecular Chemistry Symposium on "Tailormade Polymers", Merseburg, March 20–21, 2000.

The national Symposium of the Division of Macromolecular Chemistry of the GDCh (Fachgruppe Makromolekulare Chemie der Gesellschaft Deutscher Chemiker) in 2000 was held in Merseburg with a topic of *Tailormade Polymers* supplemented by a poster session on *Contemporary Polymer Chemistry.* This was after Dresden, 1996, and Mainz, 1998, the third choice of a new location after the Division has decided that the former traditional *Bad Nauheim* Symposium will be held in different polymer centers in Germany.
After the big success of the proceeding Symposium in Mainz, the organizers took the challenge to select the smaller city Merseburg as conference site with a strong tradition in polymer science and technology and situated in the so-called former Chemistry Triangle. A special poster session and an opening session were dedicated to the activities in polymer science at the chosen conference site. Thus, the prorector of the University Halle-Wittenberg, Prof. Janowski, the rector of the Fachhochschule Merseburg, Mrs. Prof. Wanka, and the major of Merseburg, Mr. Glietsch, welcomed the participants and gave an excellent information on the history, the present status, and the activities in Merseburg as traditional center for Polymer Science. Mrs. Prof. Sandner provided details on the research activities in the Institute of Macromolecular Chemistry of the University Halle-Wittenberg. The opening session was concluded with the handing over of the year 2000 Reimund Stadler awards of the Division Macromolecular Chemistry.
The scientific program with the general topic of "Tailormade Polymers" was divided in two parts: contemporary activities in polymer synthesis and the tailoring of polymer properties by suitable modification steps. These fields were introduced by four excellent plenary lectures and further deepened in 37 shorter lectures presented in two parallel sessions. All modern synthetic techniques for controlling structure and properties were represented. Consequently, polymer architectures like block copolymers, hyperbranched polymers and also nano tubes achieved by these techniques, played also an important role. Not less interesting were the contributions on tailoring polymer properties by modification. The lectures covered surface modification of particles, membranes, biomaterials, and saccharides. A large section was focused on polymer blends, fiber composites, and in general polymer modification in the melt. The final highlight of the German Division Macromolecular Chemistry Symposium was the international plenary lecture presented by Dave Tirrell on Biological Synthesis.
An excellent insight into contemporary activities in polymer synthesis, modification and characterization was given by the large poster sessions with in total 128 poster presentations from groups all over Germany.
Without any doubt this meeting was again a big success and we would like to thank all contributors and the participants for their engagement. 419 scientists and students from academia and industry came to Merseburg to participate in this two-day symposium. It was good to see that the German polymer community accepts this symposium which is somewhat more focused on industrial aspects

than e.g. the Freiburg Makromolekulare Kolloquium. A selection of some of the contribution is given in this Symposium Volume.

*Brigitte Voit*
*Institute of Polymer Research Dresden*

# Controlled Polymer Synthesis in Homogeneous and Heterogeneous Processes

**Markus Klapper and Klaus Müllen**

**Max-Planck-Institut für Polymerforschung**

**Ackermannweg 10, D-55128 Mainz**

**Germany**

## Abstract

Research in polymer synthesis can be oriented toward method or structure. This review is intended to document both approaches. Recent examples from our own work as well as future prospectives will be presented. A general concern is to work out the impact from organic and supramolecular chemistry.
Examples of method oriented work are (i) a unique polycondensation via carbon sulfur-bond formation providing hybrid structures of, e.g., polyphenylenesulfide and polyaniline; (ii) a triazoline additive establishing a novel self-regulation process in controlled radical polymerization; (iii) a metallocene catalyzed polyolefin synthesis which is performed by using "smart", reversibly cross-linked polymer resins as support; and (iv) hydrogen-bonded aggregates suggesting new emulsifiers for emulsion polymerization thus leading to functional nanoparticles.
Examples of structure oriented work are (i) rod-coil block copolymers (1D case) as a „covalently bound combination" of rigid and flexible segments whereby a key synthetic concern is the chemical modification and quantitative end-functionalization of polyphenylene chains; (ii) discs (2D case) as molecularly defined graphite subunits which are obtained via a remarkably mild intramolecular cyclodehydrogenation; and (iii) dendrimers (3D case) as shape-persistent nanoparticles whose high degree of structural precision results from a new cycloaddition-deprotection protocol.

## 1. Introduction

Polymer synthesis is a multi-tasking process since it (i) focusses on both structures and methods; (ii) tries to control molecular and supramolecular design, at different length scales and different phases, and (iii) has a strong bias for the active function of polymers, i.e., on their role as materials. With this in mind the outline of the present text appears straightforward: We take a look into quite classical synthetic methods, polycondensation (section 2.1) and radical polymerization of activated olefins (section 2.2), thereby going for

 CCC 1022-1360/00/$ 17.50+.50/0

new ways of repetitive bond-formation and improved control of radical polymerization. While both methods of polymer synthesis can be discussed by restricting the attention to isolated molecules in solution, the following approaches, polyolefin synthesis using supported metallocene catalysts (section 2.3) and emulsion polymerization (section 2.4), have an additional supramolecular aspect since particle formation and bonding through non-covalent effects come into play as well. Needless to say, a spatial confinement of the polymer forming reactions has a profound influence on the synthetic concepts to be applied.

In the search for new, unconventional polymer structures, our starting point will be a classical example, that of polyphenylenes. The dimensionality of polyphenylenes will serve as a guideline, since we proceed from the well known rigid rods, the 1D case (section 3.1), to giant discs, the 2D case (section 3.2), and to dendrimers, the 3D case (section 3.3). Dendrimers which are constructed only from benzene moieties differ significantly from already known dendrimers and can serve as unique functional nanoparticles. It is the astonishing aspect of our design process that a plethora of unique molecular and supramolecular architectures is derived from a seemingly simple chemistry using the benzene ring as modular building block.

The following article is intended to outline our recent work in polymer synthesis aiming at both new synthetic methods and new polymer structures. For details the reader is urged to consult the original publications since this overview is necessarily written in a concise and compact manner. This text is designed to cover quite different topics which at a first glance might even appear somewhat unrelated. These cases are meant to demonstrate, however, how the general needs of controlled polymer synthesis can be met for a diversity of methods and structures. The other advantage of this approach is that it stimulates a fruitful and sometimes surprising combinations of expertise.

## 2. Methods

### 2.1. Polycondensation by Carbon-Sulfur Bond Formation

Are there still challenges in polycondensation ? Let us invoke two classical examples of polycondensation products: poly(phenylenesulfide) (PPS, 1) and polyaniline (PANI, 2). Both polymers are obtained by easy procedures and have great practical importance, but suffer from a series of drawbacks, such as low solubility and limited tractability. We thought that we to make a hybrid structure by combining the structural features of PPS (**1**) and PANI (**2**) (*Scheme 1*). [1]

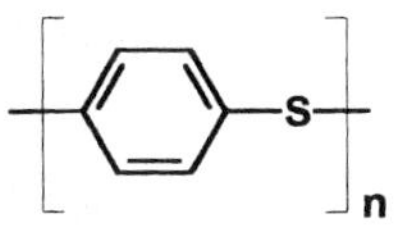

PPS (1)

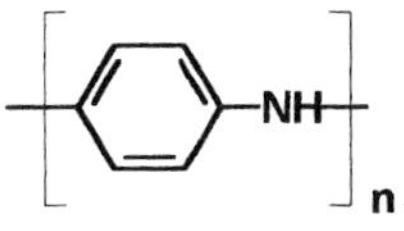

PANI (2)

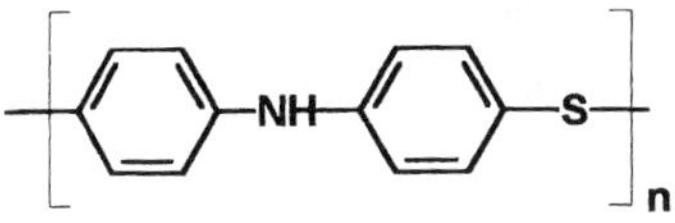

PPSA (3)

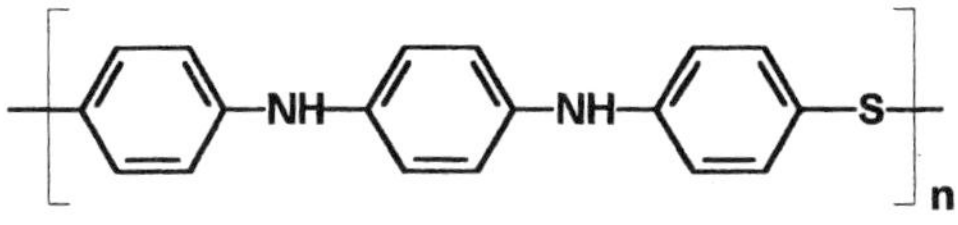

PPSAA (4)

***Scheme 1:*** Sulfur and nitrogen-bridged polyphenylenes

Our synthetic concept towards the so-called poly(phenylenesulfide-phenyleneamine) (PPSA, 3) is based on an acid-induced condensation of aromatic methylsulfoxides. Upon treatment of the AB-type monomer 5 with strong acids, polysulfonium salt 6 can be obtained in high yields. Demethylation of 6 with pyridine as the nucleophile gives nearly quantitative access to high-molecular weight PPSA (3) (Mn up to 150.000) *(Scheme 2)*.

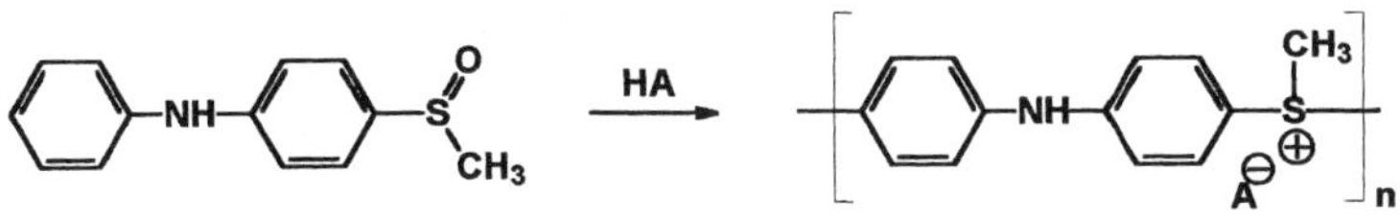


5 6

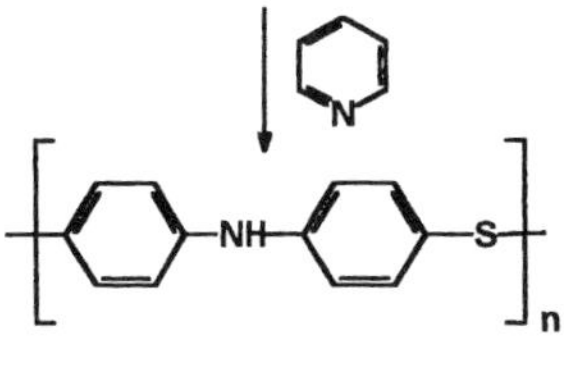


3

***Scheme 2:*** Synthesis of PPSA

In contrast to the classical polymers PPS and PANI, PPSA is structurally well-defined and exhibits exceptional solubility in organic solvents, such as THF or DMF. Flexible, transparent and colorless films can easily be prepared by spin-coating techniques. Oxidative doping of these films with iron(III)chloride results in electrical conductivities of about 10 S/cm, which remain constant over several months. Apart from promising conductivity, PPSA has turned out to be a very efficient hole-transport material for LED-devices, which, together with its excellent solution processing, defines the material as a prominent candidate for this special application. [2]

The above concept for carbon-sulfur bond formation was extended towards the synthesis of high-molecular weight poly(phenylenesulfide-phenyleneamine-phenyleneamine) (PPSAA, 4), a structure which is attractive due to its close electronic relationship to PANI 3. [3] Chemical oxidation of the as-synthesized, completely reduced form of PPSAA (4) results in different oxidation states of the polymer 7, analogous to PANI (*Scheme 3*).

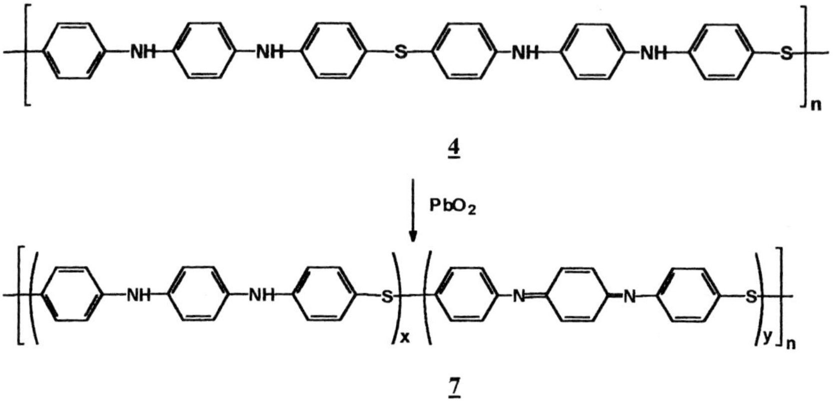


***Scheme 3:*** Oxidation of PPSAA

Remarkably, PPSAA (7) is soluble in solvents such as THF and NMP through all different oxidation states, and flexible, free-standing films can easily be prepared. Furthermore, by addition of any type of acid to the half oxidized "emeraldine-like" PPSAA ($x = y = 0.5$), spontaneous electron transfer and the formation of semiquinone species are observed. Hence, an acid-type doping without changing the number of $\pi$-electrons similiar to PANI is possible.

We conclude that a new method of bond formation gives access to high molecular weight polymers without structural defects and allows control of the physical properties of the products. Further opportunities can be highlighted by the synthesis of hyperbranched analogues of PPSA and by an intramolecular version of the carbon-sulfur bond formation which provides access to sulfur containing poly[n]acenes as double-stranded polymers.

## 2.2. Self-Regulation in Radical Polymerization

In radical polymerization of activated olefins the similar question of rigorous control over the polymer structure goes back to the problem of how we avoid any reactions of the growing free macroradical except chain propagation. Side reactions such as dimerization or disproportionation will not only reduce the thermal stability of the product, but also limit the ability of the free macroradical to participate in block copolymer formation upon reaction with another monomer.

Two methods have recently been established to achieve living species in radical polymerization: the ATRP (atom transfer radical polymerisation) and the NMRP (nitroxide mediated radical polymerisation). Both processes allow the synthesis of block copolymers, but for a technical application some disadvantages have to be solved. In the case of ATRP, molar amounts of Cu catalyst have to be removed, requiring an additional purification step; and, furthermore, polar monomers cause serious difficulties due to the interaction with the metal ion. The most important disadvantage in NMRP for the most common additive TEMPO is the limitation to autopolymerizable monomers such as styrene and styrene derivatives.[4] This is a result of suppressed, but still occurring irreversible side reactions of free polymer radicals such as recombination or disproportionation. [5]

Due to the drawbacks in the existing procedures, metal content and restriction to autopolymerizable monomers, there is still a demand for new additives for controlled radical polymerization.

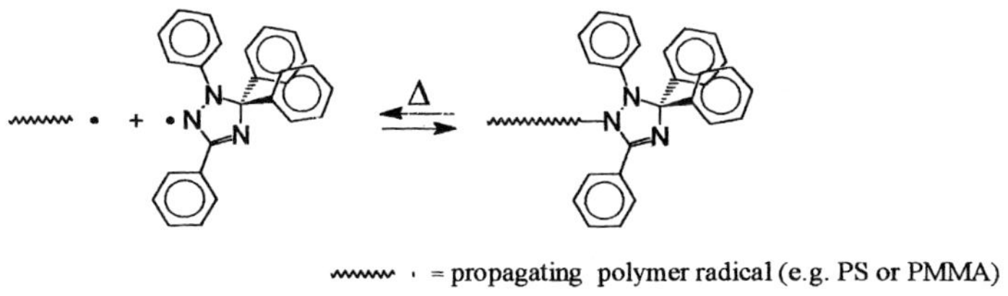


**8**

***Scheme 4:*** Equilibrium between active and dormant side with triazolinyl as counter radical

We demonstrate that triazolinyl radicals **8** (*Scheme 5*), first published by Neugebauer [6], fulfill the requirements for a more applicable system.[7] Not only styrene, but also MMA can be polymerized up to high conversions in a well controlled way. Triazolinyl radicals have some significant advantages over TEMPO. The polymerization rates are about four times higher and a broad variety of monomers can be polymerized in a controlled way. While molecular weights in the TEMPO controlled styrene polymerization are limited up to 20000, in the triazolinyl case molecular weights of PS and PMMA of around Mn > 100 000 (*Scheme 5b*) are accessible. *Scheme 5a* clearly indicates the linear increase of the molar mass versus conversion in the polymerization of styrene and in a similar way, (*Scheme 5b*) proves the controlled process by the shift of the GPC traces towards higher masses with increasing conversion.

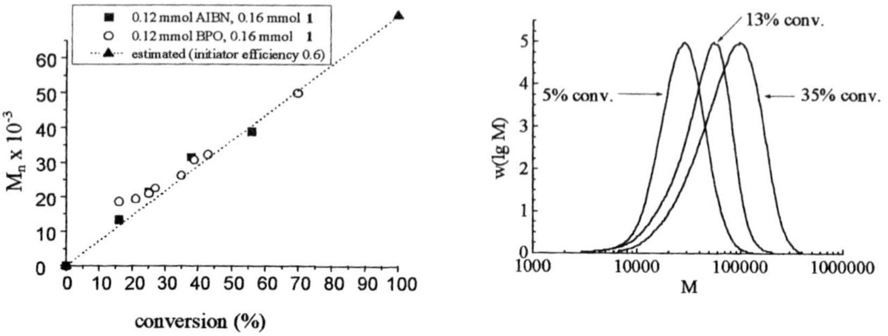


***Scheme 5:*** a) Controlled radical polymerization of styrene in presence of triazolinyl **1** (T = 120 °C, bulk); $M_n$ versus conversion; b) Shifting of the GPC traces towards higher masses with increasing conversion

Via mass spectrometry and UV-spectroscopy a high degree of endfunctionalization can be shown. This results not only in the possibility of chain extension but, more important, in the generation of new types of blockcopolymers such as PS-b-PVAc or Poly(TMS-HEMA-b-PS) (*Scheme 6*).

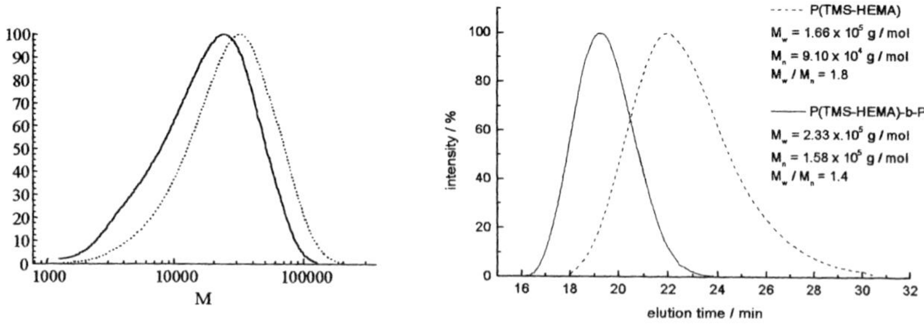


***Scheme 6*:** a) GPC traces of poly(styrol) macroinitiator versus the resulting poly(styrol)-b-PVAc; b) GPC traces of poly(TMS-HEMA) macroinitiator versus the resulting poly(TMS-HEMA)-b-polystyrene

To explain the ability of this radical to control a broad variety of monomers one has to take a closer look into the kinetics of the polymerization. In the TEMPO case it was proven that due to side reactions the number of polymer chains which possess the ability to undergo further chain growth steadily decreases. In order to maintain the equilibrium constant K=([P•][T•])/[P-T] invariant (*Scheme 7*), this loss of „living“ polymer chains is balanced by a simultaneous increase of the number of persistent counter radicals T•, subsequently leading to a shift of the polymerization equilibrium to the dormant side.

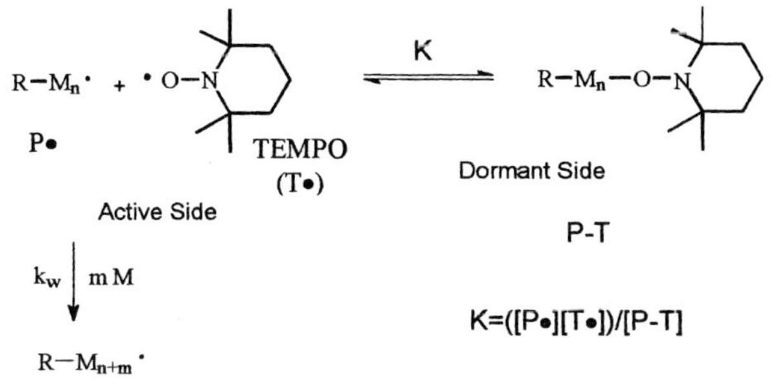


*Scheme 7:* Equilibrium between active and dormant side with TEMPO as additive

This unwanted situation is avoided when the monomer is susceptible to spontaneous (thermal) initiation as it is the case for styrene and styrene derivatives. The thus newly formed radical chains capture the excess of free counter radicals.

However, in our case not only autopolymerization can generate new growing radicals, but also the triazolinyl radical itself is able to control the amount of macroradicals. This surprising result is an effect of a slow decomposition (half lifetime 20 min / 100°C) of the free radical **8** which is proven by ESR spectroscopy and FD mass spectrometry. The driving force of this reaction which gives rise to a new phenyl radical is the simultaneous formation of an aromatic heterocycle (*Scheme 8*).

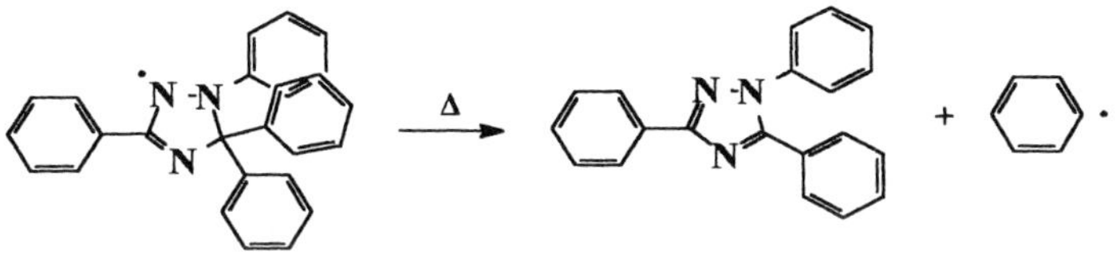


**8** **9** **10**

*Scheme 8:* Decomposition of **8**

The phenyl radical **10** strongly influences the polymerisation since it initiates new polymer chains. It is important to mention that a triazolinyl group attached to the polymer chain is completely stable because decomposition would not lead to an aromatic heterocycle. In a typical radical polymerization the period during which the counter radical is free is very short and its decomposition is nearly excluded. Decomposition occurs only if an excess of counter radical due to the side reactions.

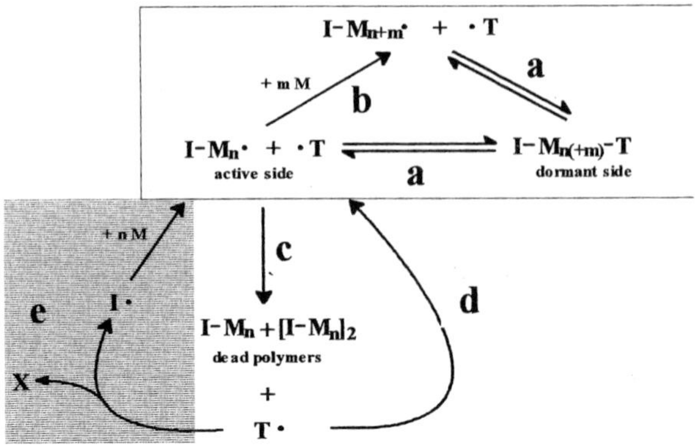


***Scheme 9:*** Concept of self-regulation

When the number of irreversibly terminated chains increases, the number of free counter radicals **T**· increases as well (**c**). Consequently, the tendency of the counter radical to decompose to **X** by release of new initiating species **I** increases. Half of the amount of the triazolinyl decomposes and forms an initiator (**e**), the other half acts as controlling agent for these newly formed chains (**d**). Accordingly, the counter radical itself possesses the possibility to control its free concentration and therefore the equilibrium **a** (*Scheme 9*). [8]

In conclusion, with the triazolinyl radicals, due to a special mechanism, a broad variety of monomers can be polymerized in a controlled, metall free way, autopolymerization like in the case of TEMPO is no longer essential since the system controls itself by the side reactions. In the future, this additives will allow for the synthesis of block copolymers with unusual monomer combinations. Even water-soluble or very polar monomers can be easily incorporated resulting in materials with special morphologies in bulk or agreggation behavior in solution. Especially, for using polymers as emulsifiers, the easy synthesis of amphiphilic block copolymers will be a challenge.

## 2.3. Supported Metallocenes in Polyolefin Synthesis

It has been emphasized already that the following example of polymer synthesis is different in that it implies a heterogeneous process. The key step is the synthesis of a „smart“ polymer resin which acts as support for catalyst systems. Here again, however, the novelty comes from a concept of organic chemistry since the cross-linking is achieved by a reversible Diels-Alder cycloaddition.

The synthesis of polyolefins is by far the most important industrial polymer-forming process. [9] Compared to conventional Ziegler-Natta catalysts, the metallocene/aluminoxane catalysts offer a number of advantages. In particular, they combine high activity with the possibility to tailor polymer properties such as molecular weight, molecular weight distribution, co-monomer insertion and distribution, as well as the stereoregularity of α-olefins.
Hitherto, most articles have dealt with the modification of the metallocene ligand sphere in order to generate a series of efficient, soluble metallocene catalysts. [10] For industrial gas-phase and slurry polymerization processes, however, heterogeneous catalyst systems are an important requirement. [11] The catalyst support acts as a template in these particle growth processes and controls the size and morphology of the resulting polymer. Commonly used inorganic supports for metallocenes, such as $SiO_2$, $Al_2O_3$ or $MgCl_2$, show a rigid and polar surface structure as opposed to polymer carriers. Hydrocarbon support materials provide a much closer analog to the environment prevailing in homogeneous polymerization. [12, 13] In contrast to polystyrene beads cross-linked with divinylbenzene, where all synthetic steps are performed heterogeneously and only on the accessible surface, our approach involves the formation of the resin following the attachment of metallocenes to the polymer backbone.

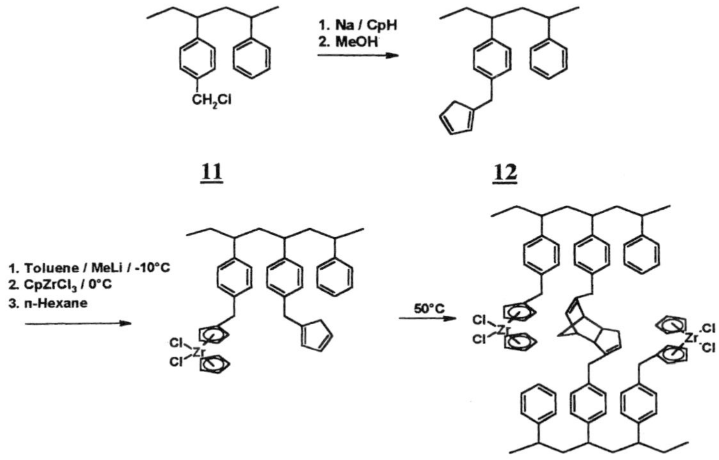


13 14

***Scheme 10:*** Synthesis of a metallocene catalyst on a reversibly cross-linked polystyrene

This is expected to provide a more uniform distribution of active centers within the particles (***Scheme 10***). [14] Furthermore, the reversibility of the cross-linking Diels-Alder reaction allows for fragmentation of the catalyst support during the polymerization and polymer processing. The swellable and reversibly cross-linked polymer support which is introduced here, thus, constitutes a closer analog to the soluble metallocene system. It allows for seemingly heterogeneous polymerization to proceed in an isotropic, dynamic and microscopically solution-like environment. [15] This is in sharp contrast to the situation prevailing in rigid, anisotropic surfaces of inorganic supports.

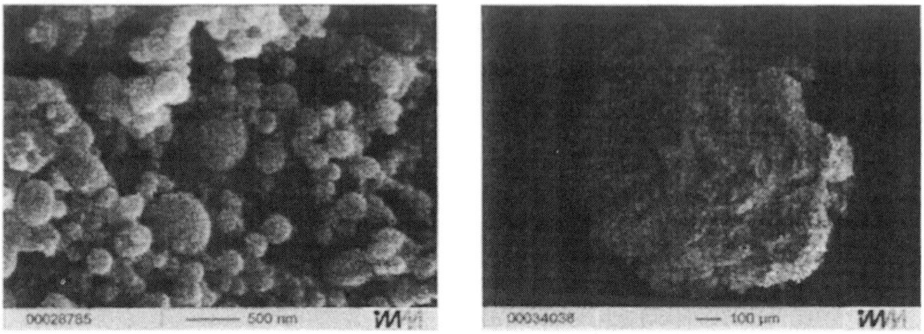


***Scheme 11:*** SEM microscope image of a) a metallated particle (scale bar 500 nm)
b) of a product bead (scale bar 100 µm)

Remarkably enough, despite the easy synthesis of our catalyst the polyolefin product is isolated as nearly spherical beads of millimeter dimension as proven by scanning electron microscopy (*Scheme 11*).

It is quite obvious that this synthetic approach is restricted to cyclopentadienyl containg catalysts. However, to control tacticity in polypropylenes more complicated catalysts have to be applied. Therefore it is not very reasonable to generate them on a polymeric backbone which would require a multi-step synthesis but to synthesize them separetely and combine them on the final stage with the polymeric support.

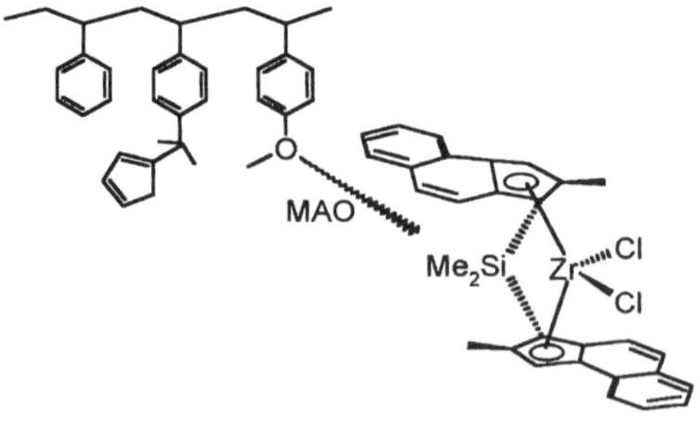


***Scheme 12:*** Non-covalently attached metallocene

The synthetic approach for such a support is quite straightforward since one starts with a copolymer of styrene, bromostyrene and methoxystyrene, lithiation and addition of fulvene results in a cyclopentadienyl containing polymer. Again, the cyclopentadienyl unit is responsible for the crosslinking. [16]

For attaching the catalysts the methoxy group is essential. Due to the high affinity of MAO towards ether groups and to the metallocenes, the catalyst can be non-covalently, but with a high stability, attached on polymeric materials via a "MAO-bridge".

This concept allows, similar to the industrial process using silica, the supporting of a nearly unlimited number of metallocene catalysts on a polymeric material. Polypropylenes with different tacticity can be generated and, furthermore, activity and morphology of product particles optimzed depending on the degree of cross-linking, loading of metallocene and amount of activator. The extension towards copolymerisation with polar monomers is a further task to obtain materials acting as compatibilizers or adhesives.

## 2.4. Hydrogen-Bonding and Emulsion Polymerization

The formation of particles which is crucial for the above approach toward supported catalysts results from cross-linking of polymer chains via covalent bonding. This suggests an alternative concept for toward the formation of networks and of particles, that via non-covalent bonding.

5-Alkoxyisophthalic acids ($C_n$ ISA) *(Scheme 10)* are challenging building blocks for self-assembly processes since they combine two powerful structure-forming principles: alkyl chain packing and hydrogen bonding.

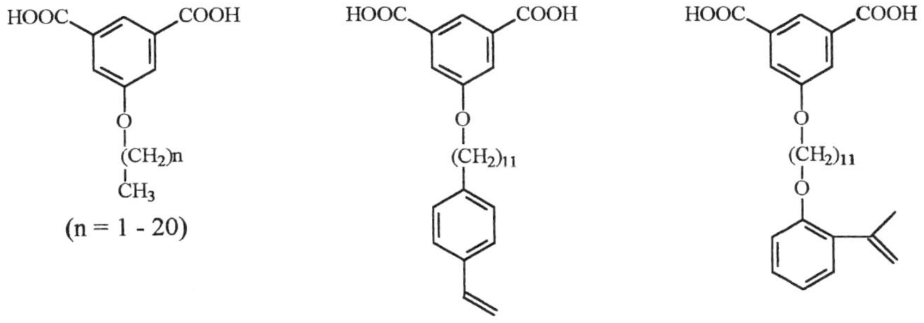


15 16 17

***Scheme 13:*** 5-Alkoxyisophthalic acid ($C_n$ISA) (15), 5-((4-styryl)undecyl)oxy) isophthalic acid (ISA-$C_{11}$-Sty) (16) and 5-((2-propenylphenoxy)undecyl)oxy) isophthalic acid (ISA-$C_{11}$-ipr) (17)

When investigating their packing behavior as a function of the alkyl chain length, crystal structure analysis reveals that in the diacids $C_n$ISA, with n < 6 or n > 12, the alkyl substituents from adjacent molecules of hydrogen bonded strands point towards opposite directions and that infinite sheets with interdigitating alkyl chains are formed (*Scheme 14*). [17] In contrast, the $C_n$ ISA molecules with 6 < n < 10 form cyclic hexamers in the crystal. [14] Furthermore, the number of motifs in the crystalline lattice can be extended by adding difunctional aromatic bases such as pyrimidine or pyridazine. Here again, strong hydrogen bonded arrays are seen throughout the lattice. An induction of mesophases is also possible.[15] 5-Alkoxyisophthalic acids also form highly ordered monolayers on graphite upon physisorption from solution which can be visualized in real space by scanning tunneling microscopy. This allows a systematic comparison of pattern formation in 2D- and 3D-crystals and paves the way toward chemical transformations of single molecules.

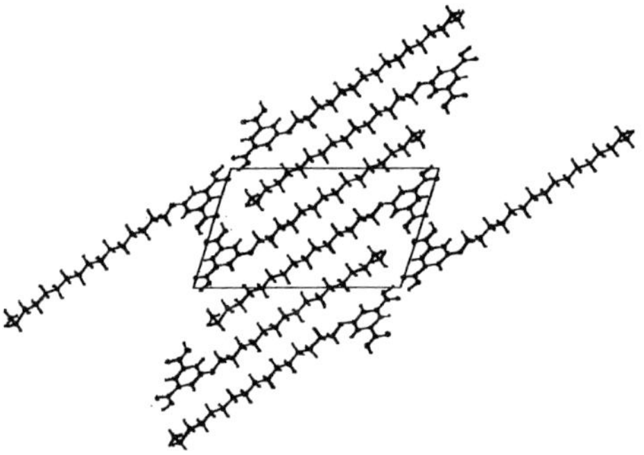

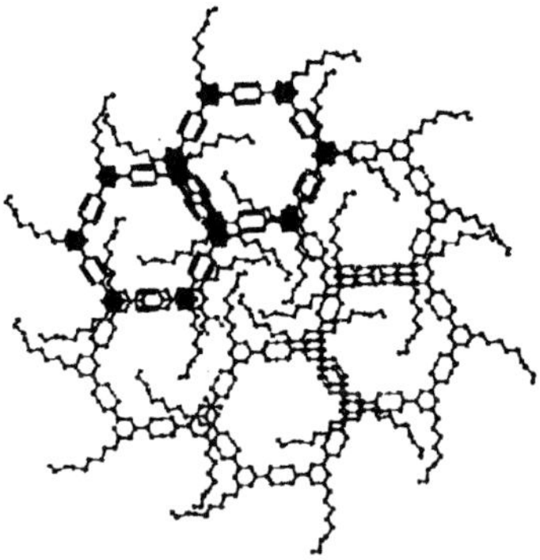

***Scheme 14:*** Crystal structures of $C_{16}$ISA (**15**) Channel-type crystal structure of $C_8$ISA

Moreover, the amphiphilic character of the alkoxyisophthalic acid derivatives makes it possible to utilize them as surfactants in emulsion polymerization of different monomers (e.g.: styrene, n-butylmethacrylate and n-butylacrylate) (*Scheme 15*). Indeed, such dicarboxylic acid surfactants increase the electrostatic stabilization of the final latex. The CMC (critical micellar concentration) of these compounds is comparable with the CMC of other dicarboxylic acids [18, 19]. Conventional amphiphilic surfactants used in emulsion polymerization stabilize the final latex particles by adsorption at the polymer/water interface. The major problem of these surfactants appears during the handling of the dispersions. The surfactant is able to adsorb or desorb freely from the latex surface affecting the latex stability and mechanical properties.

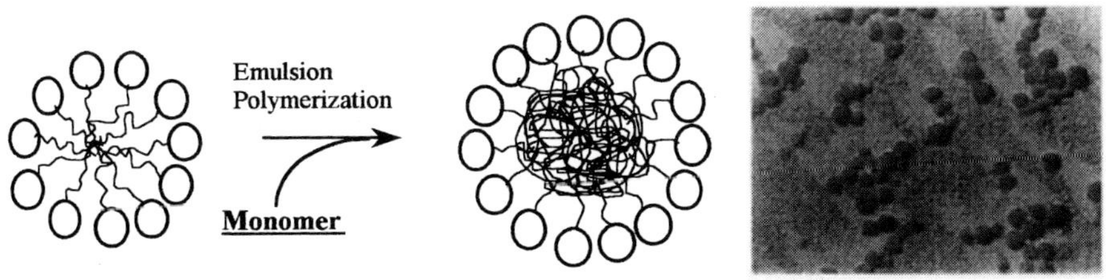


***Scheme 15:*** Emulsion polymerisation in the presence of **15** (Monomer: styrene, TEM picture)

One possibility to improve latex stability is the copolymerization with polymerizable surfactants (surfmers). We synthesized different surfmers and copolymerized them with styrene. The $^1$H NMR study of the monomer-surfmer conversion ratio showed that ISA- $C_{11}$-sty (**16**) and the ISA-$C_{11}$-i-pr (**17**) are copolymerized with styrene. The final polymer dispersions are monodisperse and very stable in the presence of mono- and divalent salts (e.g., KCl and $CaCl_2$).

A potential application results from the possibility of functionalizing the latex particles in aqueous media with functional components such as oligopeptides or dyes *(Scheme 13)*. As a result, the carboxylate groups on the surface of the copolymer lattices are involved in the formation of covalent bonds with the functional units. The great number of opportunities for assay formation is obvious. A further challenge comes from a comparison of latices and dendrimers (see below) as functional nanoparticles of different size and stability.

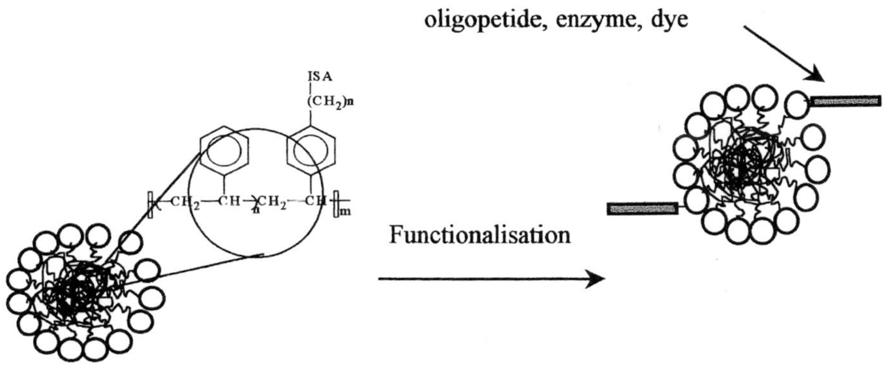


***Scheme 16:*** Functionalisation of latex particles

# 3. Structures

## 3.1 Rigid-Rod Polyphenylenes and Rod-Coil Block Copolymers

We have in section 2 dealt with important classical and practical problems of polymer synthesis in both homogeneous and heterogeneous processes and made attempts at introducing new solutions to these problems with a strong influence from organic and supramolecular chemistry.

As mentioned in the introduction, our search for new polymer structures is restricted to polyphenylenes of varying dimensionality. Poly-para-phenylene is the classical case of a rigid-rod and the prototype of a conjugated polymer. Conjugated polymers are of great academic as well as industrial interest since they possess a large application potential [20, 21].

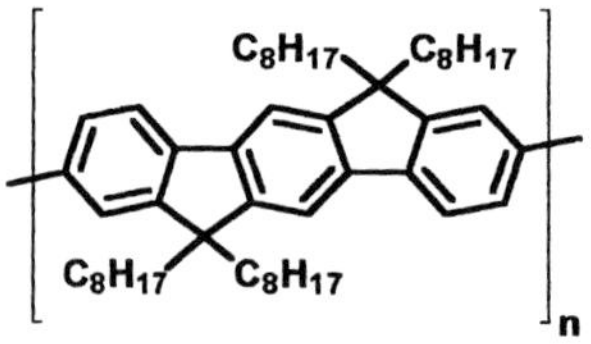


**18**

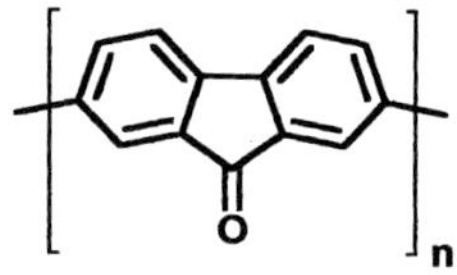


**19**

*Scheme 17:* Examples of polyfluorene-type structures

Recently, our efforts have focussed on bridging the gap between the step-ladder type poly(2,7-(9,9-dialkylfluorene)) and a true ladder-type polyphenylene by the synthesis of poly(2,8-(6,6'-12,12'-tetraalkylindenofluorene)) (2,8-PIF) (**18**) which contains three planarized phenylene rings as repeating unit. [22] Polymer **18** exhibits the desired bathochromic shift of absorption and emission bands with respect to the polyfluorenes (**19**). Further, the material posseses excellent film-forming properties, is stable up to 400 °C and forms liquid-crystalline phases.

The device performance of a light emitting diode (LED) depends sensitively upon the balanced injection of holes and electrons into the active layer; therefore, hole-transport as well as electron-transport materials have been used in multi-layer devices. Our efforts towards the development of novel, efficient electron transport materials led to the preparation of poly-(2,7-fluorenone) (**19**) via a precursor route. [23] Cyclic voltammetry measurements revealed a reversible reduction with a $E_{1/2}^0$ of -1.48 V vs $Fc/Fc^+$ which makes the polymer a promising candidate for the use as the electron transport layer in multi-layer LEDs.

A novel strategy towards controlling the self-assembly, film morphology and hence the opto-electronic properties of conjugated polymers invokes the synthesis of luminescent rod-coil block copolymers. Even short block lengths lead to phase-separated structures in the nanometer regime. [24] The block ratio of the luminescent rods and the non-emissive coils plays an important role, since on the one hand the block lengths should be long enough to seggregate, and on the other hand the length of the flexible coil should be minimized to obtain a material with the highest possible concentration of the conjugated species.

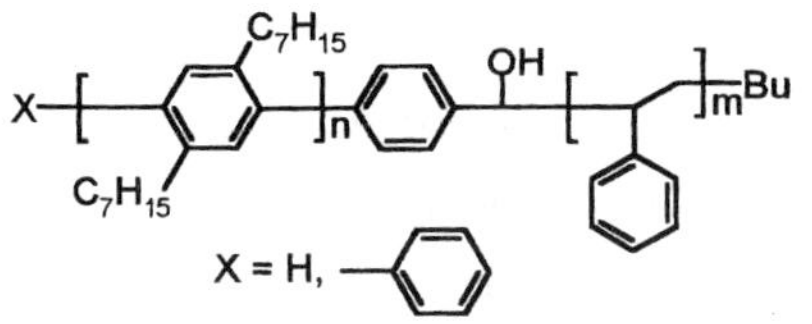

20

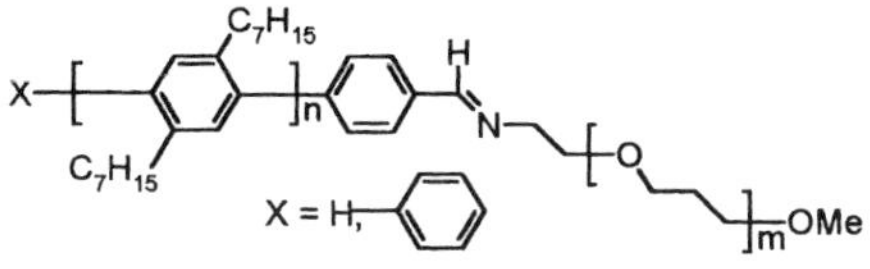

21

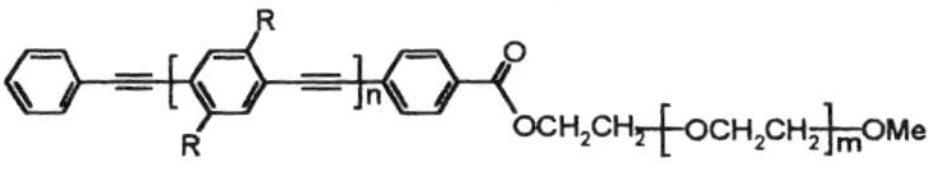

22

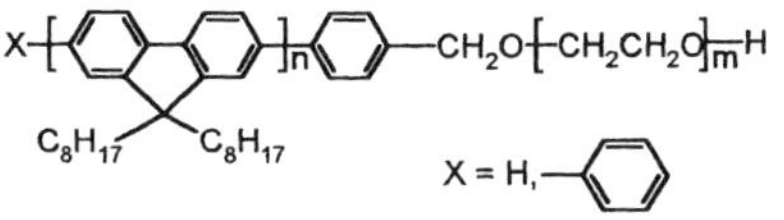

23

***Scheme 18:*** Some examples of rod-coil blockcopolymers

Synthetic strategies towards rod-coil block copolymers involve either a condensation of two pre-formed blocks (grafting-onto process) or polymerization using a macroinitiator (grafting-from process). A quantitative end-functionalization of the desired luminescent rod polymers is a prerequisite for both approaches. We were able to perfectly end-functionalize polyphenylenes by using transition-metal mediated coupling reactions. [25, 26] End-functionalized polyphenylenes (PPPs) and polyphenyleneethynylenes (PPEs) rods were in a second step used as building blocks in conventional condensation protocols followed by careful work-up procedures to obtain the desired rod-coil block copolymers **20 - 23** in good yields. By using a benzylalcohol-terminated poly(2,7-(9,9-dioctylfluorene)) as a macroinitiator for the polymerization of ethyleneoxide we were also able to synthesize the rod-coil block copolymer **23** by a grafting-from procedure (*Scheme 18*).

All the block copolymers show a variety of different phase-separated structures which are dependent on the film preparation. In the case of the poly(2,7-fluorene)-*b*-poly(ethyleneoxide) **23** we could even show a significant change in the optical properties of the block copolymer compared to the corresponding homo-polyfluorene in the solid state due to ordering phenomena. While morphologies of coil-coil block copolymers have long been attracted the attention of polymer scientists, the corresponding rod-coil structures are much less common. Accordingly, it will be the challenge of future synthetic work to provide a broader variety of rod-coil blockcopolymers with different natures and lengths of both components.

## 3.2. Molecular Graphite Subunits

The chain stiffness of the above rod-like polymers has already served as a powerful supramolecular principle. We now introduce a related approach for 2D- and 3D-molecules. With new molecular topologies of polyphenylenes in hand, the self-assembly of discs and spheres will then create a versatile route to functional nanoparticles.
A situation which is electronically different from that in linear polyphenylenes arises if the molecules possess two translational periods such as in disc-type polycyclic aromatic hydrocarbons (PAHs). Looking to capitalize on the interesting electronic properties of graphite, we synthesized topologically distinct graphite-like molecules. Examples of such molecules include hexa-*peri*-hexabenzocoronene (HBC) **24**, supernaphthalene **25** and supertriphenylene **26** *(Scheme 15).* [27]

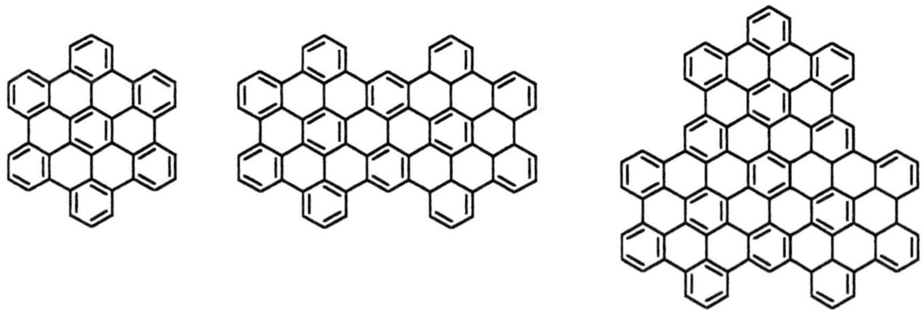

**24** **25** **26**

***Scheme 19:*** Graphite-like molecules

Conceptually, our synthetic method involves synthesis of a branched polyphenylene as precursor, followed by the planarization of the molecule to yield the final two-dimensional product. An important synthetic step to these precursor polyphenylenes is the cyclotrimerization of suitably substituted diarylacetylenes (see *Scheme 16),* but Diels-Alder cycloadditions between tetraarylcyclopentadienones and diarylacetylenes can be used as well. The construction of polyphenylenes is then followed by their oxidative cyclodehydrogenation and planarization into the final „graphite sheets“. [28]

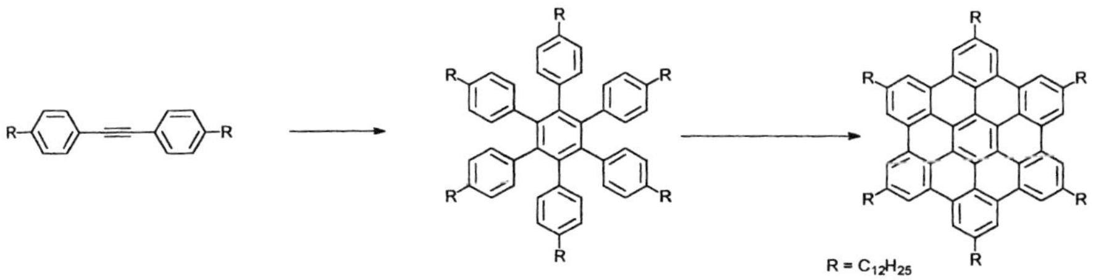


**27** **28** **29**

*Scheme 20:* Synthesis of hexa-*peri*-hexabenzocoronene (HBC)

Another example is molecule **30** which after cyclotrimerization and planarization gives the PAH **32**. Characteristic for this large polyaromatic hydrocarbon is the π-stacking between adjacent layers with insolubility as a consequence.

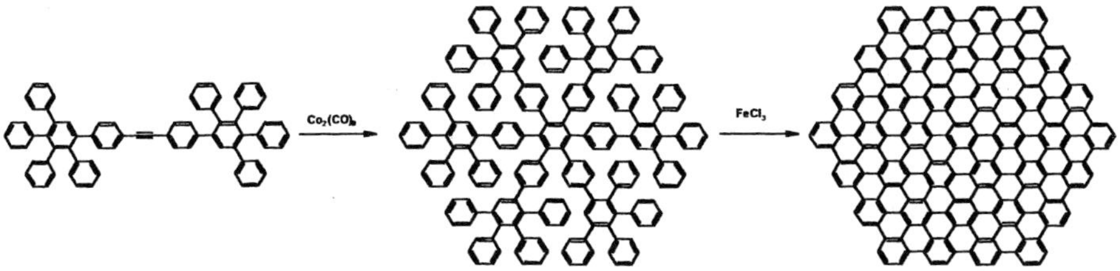


**30** **31** **32**

*Scheme 21:* Synthesis of the $C_{222}$-PAH **32**

Unfortunately, it is difficult to fully study the electronic properties of these unsubstituted molecules due to their insolubility. To overcome such problems, we have also synthesized their alkyl substituted derivatives. Interestingly, HBCs such as **19** form thermotropic columnar hexagonally ordered mesophases ($Col_{ho}$) at higher temperatures. The $Col_{ho}$ mesophase involves the specific arrangement of the disc-like aromatic HBC cores, packed co-facially to optimize π-π interactions of the aromatic cores, into columns and these columns are arranged in a hexagonally close-packed fashion. Typical HBC derivatives are crystalline at room temperature and enter the mesophase at about 110°C. Isotropization temperatures are in excess of 420° C. These mesophases have remarkably large phase widths in comparison to columnar mesophases obtained from conventional disc molecules and are of interest because of their one-dimensional charge-carrier transport properties. To date, these HBC mesophases exhibit world record charge-carrier mobilities of the columnar-type mesophases.[29]

In order to preserve the hexagonally-packed columnar supramolecular ordering down to room temperature, we created cross-linkable HBC derivatives with acrylate and methacrylate functions at the terminus of each alkyl chain. These molecules could be easily thermally cross-linked and indeed, the $Col_{ho}$ arrangement was preserved to room temperature with no distortion of the molecular packing.

To tailor the mesophase behavior (e.g. via H-bonding), mono and difunctionalized HBC molecules were desired. A more flexible route to a variety of substituted HBCs is the above mentioned coupling of diarylacetylenes to functionalized cyclopentadienones.[30] The bromo-substituted compounds can be transformed into a variety of functional groups (e.g. amino, carboxylic acid, cyano), which can affect both the supramolecular packing and the electrochemical potential of the molecule. [31, 32] A potent example of substituents affecting mesophase behavior is the case of HBC carrying six 4-alkylphenyl substituents (HBC-$PhC_{12}$). The $Col_{ho}$ mesophase of HBC-$PhC_{12}$ exhibits both better packing of each HBC core within one column and longer range hexagonal packing of the columns, as seen in WAXS and NMR spectroscopy, when compared with the standard n-alkyl HBC derivative. [33] Accordingly, the control of the supramolecular order and of the resulting physical properties is a synthesis driven process since in the design of the molecules one includes structural features relevant for the packing behavior.

By making molecular graphene sheets increasingly larger, polymer synthesis approaches the world of carbon-rich materials in which, e.g., graphitic or tubular structures are produced by unselective high-temperature methods. As part of this merger of different fields we are introducing a precursor route in which graphene sheets are made on substrate surfaces with the latter serving as chemical template.[34]

## 3.3 Dendritic Polyphenylenes

In section 3.2 branched polyphenylenes have served as precursors for fully conjugated PAHs. Surprisingly enough, the principles used in their synthesis also lead to a new type of dendritic 3D-polyphenylenes.

We have recently described a rapid access to hyperbranched and dendritic polyphenylenes that contain pentaphenylbenzene repeating units and can be synthesized in high yields by a repetitive Diels-Alder cycloaddition. [30] It is the beauty of this synthetic method that even high molecular weight polyphenylene dendrimers can be obtained as monodisperse species without any structural defects. Further, by the use of different ethynyl-substituted aromatic cores **33** and **34** and ethynyl-substituted cyclopentadienones as branching monomers, **35** and

36, we were able to obtain dendritic 3D-polyphenylenes of different shapes and densities as well as with diameters varying between 2 and 8 nm (*Scheme 22*). To conclude the relevant results, the 3D-polyphenylenes

- are colorless, chemically inert and thermally stable (> 450°C),
- have good solubility in organic solvents, but are easily precipitatable in polar solvents,
- can be thoroughly spectroscopically characterized, and
- are shape persistent.

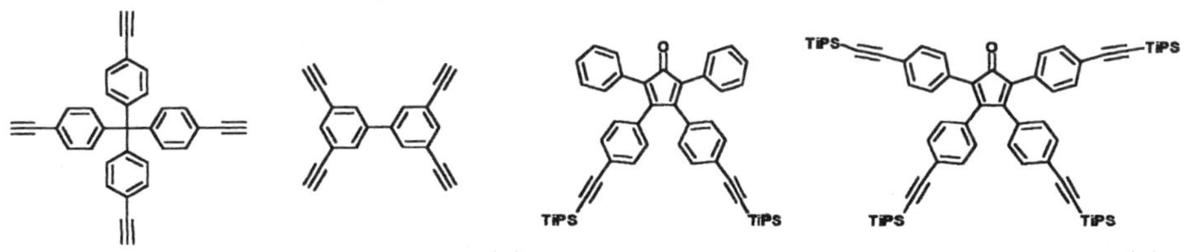


33 34 35 36

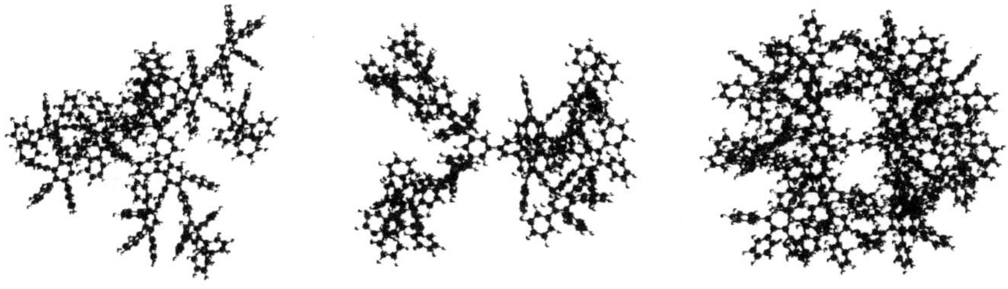

37 38 39

***Scheme 22:*** Cores, building units and the second generation dendrimers resulting therefrom

While a detailed discussion of physical properties is beyond the scope of the present text it should be mentioned that the new dendrimers are particularly rewarding objects for the visualization and manipulation of single nanomolecules by, e.g., electron microscopy and atomic force microscopy. The latter methods also support the description of the polyphenylene dendrimers as stiff 3D-particles.

Because of their stiff and dense structure, the branches of the polyphenylene dendrimers cannot interpenetrate, as has been seen from crystal structures of first generation dendrimers. [35] The pores between the dendrimer molecules are larger than 10 nm and can be filled with solvent molecules. This result is important with regard to catalysis, since many catalytic reactions are diffusion controlled and strongly dependent on pore size.

By *in-situ* deprotection and polymerization of **35**, it is possible to obtain polydisperse hyperbranched polyphenylenes with high molecular weights up to $M_w = 96000\ gmol^{-1}$, but with less well-defined shapes.

For catalysis, carrier properties and supramolecular chemistry, the dendrimers can be chemically functionalized. The first approach is simple, but unselective. By sulfonation reactions it is possible to obtain a water-soluble dendrimer. The number of attached sulfonic groups depends on the molar ratio of sulfonic acid and the dendrimer used. Chloromethyl groups are very important substituents for attaching catalysts or as initiators for polymerization. Due to electronic as well as steric reasons, by chloromethylation one can attach up to 60 chloromethyl groups on a second generation dendrimer, thus, yielding a randomly surface-functionalized nanoparticle that allows further manipulations.

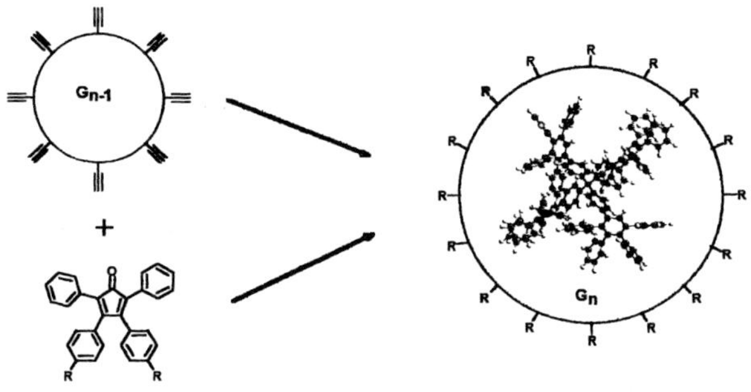


**40**

***Scheme 23:*** Functionalization of polyphenylene dendrimers, cycloaddition of functionalized cyclopentadienones in the last generation

The second concept is to use the controlled functionalization of ethynyl substituted dendrimers (section 3.2) with substituted cyclopentadienones in the last step. This method allows a topologically defined functionalization of the dendrimer surface, but requires the synthesis of specifically substituted cyclopentadienones for each functional group. The substituents **R** of cyclopentadienones **40** can be halide, cyano, methoxy, thiomethoxy, methyl, N,N-dimethylamino or hydroxymethyl groups. In polymer-analogous reactions, the substituted dendrimers can then be further modified. Suitably functionalized dendrimers allow the synthesis of core-shell polymers via various grafting-from processes. The resulting particles are unique in that they possess a particularly rigid core.

Dendrimers with sulfonic or carboxylic groups are soluble in water and can, e. g., act as drug deliverers. The polyphenylene dendrimers are also well suited as catalyst carriers. The high surface-to-volume ratio allows very high catalyst densities together with easy product separation by ultrafiltration of the dendrimer after reaction.

Using dyes on the surface of the dendrimer, unique nanoemitters can be synthesized which are exciting objects for single molecule spectroscopy. Physics has recently provided methods for the direct visualization and manipulation of single molecules or aggregates consisting of few molecules. This is where polymer synthesis comes into play since future nanoscientific developments require molecular nanoobjects of well-defined size, shape and complex functionalization.

## 4. Conclusion

A typical point of issue when looking at approaches toward polymer synthesis concerns the degree of sophistication: one school, mostly with a background in organic chemistry, goes for somewhat unconventional polymer structures and for a high degree of stuctural precision, while another school, mostly with an application orientation, stresses the need for practical solutions. The present article was meant to demonstrate that these approaches are not necessarily in conflict. A good case can be made when looking at dendrimers. Polymer scientists feeling closer to "classical structures and to sound "bread-and-butter-chemistry" might tend to regard monodisperse dendrimers as somewhat exotic species requiring overly demanding synthetic methods of synthesis. When attempting to immobilize active components with chemical, physical or biological functions on suitable support systems, it is highly appropriate to compare dendrimers, hyperbranched polymers, latices and polymer resins. Along this sequence one will always relate the necessary degree of structure control to the synthetic effort and to the ultimate desired function.

## Acknowledgement:

The advances reported here would have of course been impossible without the diligent and intelligent efforts of many members, past and present, who are unfortunately are too numerous to acknowledge here. The names contributors are collected informally within the reference section.

Financial support by the Bundesministerium für Bildung und Forschung, Volkswagen-Stiftung, Fonds der Chemischen Industrie, TMR network SISITOMAS, BRITE-EURAM project OSCA, BASF AG and SONY International (Europe) GmbH is gratefully acknowledged.

## References


[1] a) L. Wang, T. Soczka-Guth, E. Havinga, K. Müllen, *Angew. Chem. Int. Ed. Engl.* **35**, 1495 (1996); b) J. Leuninger, C. Wang, T. Soczka-Guth, V. Enkelmann, T. Pakula, K. Müllen, *Macromolecules* **31**, 1720 (1998)

[2] Y. Tak, H. Bässler, J. Leuninger, K. Müllen, *J. Phys. Chem. B* **102**, 4887 (1998)

[3] J. Leuninger, J. Uebe, J. Salbeck, L. Gherghel, C. Wang, K. Müllen, *Synth. Met.* **100**, 79 (1999).

[4] a) T. Fukuda; T. Terauchi, A. Goto, Y. Tsujii; T. Miyamato, Y. Shimuzu, *Macromolecules* **29** 3050 (1996); b) I. Li, B.A. Howell, D.B. Priddy, P.B. Smith, *Polym. Prep. (Am. Chem. Soc., Div. Polymer. Chem.)* **37(1)** 612 (1996); c) B. Keoshkerian, M.K. Georges, D. Boils-Boissier,. *Macromolecules* **28**, 6381 (1995); d) C.J. Hawker, E. Elce, J. Doer, W. Volksen, T. Russel, G. Barclay, *Macromolecules* **29**, 2686 (1996)

[5] a) Y. Zhu, B.A. Howell, D.B. Priddy, *Polym. Prep. (Am. Chem. Soc., Div. Polymer. Chem.)* **38(1)**, 652 (1997); b) M. Steenbock, M.Klapper, K.Müllen, N.Pinhal, M. Hubrich, *Acta Polymer.* **47**, 276 (1996)

[6] a) F.A. Neugebauer, H. Fischer, C. Krieger, *Angew. Chem. Int. Ed. Engl.* **28**, 491 (1989); b) F.A. Neugebauer, H. Fischer, *Tetrahedron* **51**, 12883 (1995)

[7] a) D. Colombani, M. Steenbock, M. Klapper, K. Müllen, *Macromol. Rapid Commun.* **18**, 243 (1997); b) M. Steenbock, M. Klapper, K. Müllen, *Macromol. Chem. Phys*, **199**, 763 (1998); c) D. Colombani, M. Steenbock, M. Klapper, K. Müllen, M. Fischer, J. Koch, W. Paulus, *Pat.* DE 196 36 99.7

[8] M. Steenbock, M. Klapper, K. Müllen, *Acta Polymer.* **49**, 376 (1998); M. Steenbock, M. Klapper, K. Müllen, C. Bauer, M. Hubrich, *Macromolecules* **31**, 5223 (1998)

[9] a) H. H. Brintzinger, D. Fischer, R. Mühlhaupt, B. Rieger, R. Waymouth, *Angew. Chem. Int. Ed. Engl.* **34**, 1143 (1995) b) W. Kaminsky, *Macromol. Chem. Phys.* **197** 3907 (1996)

[10] a) J. Okuda, *Nachr. Chem. Tech. Lab.* **41**, 8 (1993) b) M. Aulbach, F. Küber, *Ch. i. u. Z.* **4**, 197 (1994) and references therein

[11] F. Langhauser, J. Kerth, M. Kerstin, P. Kölle, D. Lilge, P. Müller, *Angew. Makromol. Chem.* **223**, 155 (1994)

[12] a) S. C. Hong, H. T. Ban, N. Kishi, J. Jin, T. Uozumi, K. Soga, *Macromol. Chem. Phys.* **199**, 1393 (1998), b) S.C. Hong, T. Teranishi, K. Soga, *Polymer* **39**, 7153 (1998)

[13] H. Nishida, T. Uozomi, T. Arai, K. Soga, *Macromol. Chem., Rapid. Commun.* **16**, 821 (1995)

[14] e.g. a) Y. Qian, K. Hong, H. Zong, J. Huang, *Polym. Adv. Tech.* **7**, 619 (1996); b) T. Kamfjord, T. S. Wester, E. Rytter, *Macromol. Chem., Rapid. Commun.* **19**, 505 (1998)

[15] a) M. Stork, M. Koch, M. Klapper, K. Müllen, H. Gregorius, U. Rief, *Macromol. Chem. Rapid Commun.* **20**, 210 (1999); b) M. Stork, U. Wiesler, M. Klapper, K. Müllen, *PMSE*, **80**, 8 (1999); M. Stork, M. Koch, M. Klapper, K. Müllen, H. Gregorius, U. Rief, M.O. Christen, *Pat.* DE 198 219 49.0

[16] M. Koch, M. Stork, M. Klapper, K. Müllen, *Macromolecules*, submitted

[17] S. Valiyaveetil, V. Enkelmann, G. Moeßner, K. Müllen, *Macromol. Symp.* **102** 165 (1996)

[18] P. Reb, K. Margarit-Puri, M. Klapper, K. Müllen, *Macromolecules submitted*

[19] K. Shinoda, *J. Phys. Chem.* **59**, 432 (1955)

[20] D. Marsitzky, T. Brand, Y. Geerts, M. Klapper, K. Müllen, *Macromol. Chem. Rapid Commun.* **19**, 385 (1998)

[21] a) A. Kraft, A.B. Holmes, A.D. Grimsdale, *Angew. Chem Int. Ed. Engl.* **37**, 110 (1998); b) T. Skotheim, A.B. Holmes, A.D. Grimsdale, *Handbook of Conducting Polymers*,.Marcel Dekker (1998) c) D. Marsitzky, K. Müllen, *Synth. Metals* (1999), *in print*; d) K. Ziemelis, *Nature,* **393**, 619 (1998); e) C. Kallinger, M. Hillmer, A. Haugeneder, M. Perner, W. Spirkl, U. Lemmer, J. Feldmann, U. Scherf, K. Müllen, A.Gombert, V. Wittwer, *Adv. Mater.* **10**, 920 (1998); f) F. Hide, B.J. Schwartz, M.A. Diaz-Garcia, A.J. Heeger, *Synth. Metals,* **91**, 35 (1997)

[22] a) S.R. Forrest, P.E. Burrows, M.E. Thompson, *Chem. & Ind.* 1022 (1998); b) J. v. Haaren, D. Broer, *Chem. & Ind.* 1017 (1998); c) T. Jüstel, H.Nikol, C.Ronda, *Angew. Chem. Int. Ed. Engl.* **37**, 456 (1998)

[23] S. Setayesh, D. Marsitzky, K. Müllen, *Adv. Mater.* (1999) *submitted*

[24] F. Uckert, S. Setayesh, K. Müllen, *Macromolecules* (1999) *in print*

[25] G. Mao, C.K. Ober, *Acta Polym.* **48**, 405 (1997)

[26] a) V. Francke, H.J. Räder, Y. Geerts, K. Müllen, *Macromol. Rapid Commun.* **19**, 275 (1998); b) D. Marsitzky, M. Klapper, K. Müllen, *Macromolecules* **32**, 8685 (1999)

[27] V.S. Iyer, M. Wehmeier, J.-D.Brand, M.A. Keegstra, K. Müllen, *Angew. Chem. Int. Ed. Engl.* **36**, 15 (1997)

[28] P. Herwig, C.W. Kayser, K. Müllen, H.W. Spiess, *Adv. Mater.* **8**, 6 (1996)

[29] A.M. Van de Craats, J.M. Warman, K. Müllen, J.D. Brand, *Adv. Mater.* **10**, 1 (1998)

[30] a) F. Morgenroth, E. Reuter, K. Müllen *Angew. Chem. Int. Ed. Engl.* **36**, 631 (1997); b) F. Morgenroth, K. Müllen, *Tetrahedron* **53**, 15349 (1997) c) F. Morgenroth, C. Kübel, K. Müllen, *J. Mater. Chem.* **7**, 1207 (1997) d) F. Morgenroth, A.J. Berresheim, K. Müllen, *Chem. Comm.* 1139 (1998) e) U.-M. Wiesler, K. Müllen Chem. Comm., 2293 (1999)

[31] A. Stabel, P. Herwig, K. Müllen, J.P. Rabe, *Angew. Chem. Int. Ed. Engl.* **34**, 15 (1995)

[32] J.D. Brand, K. Müllen, *in preparation*

[33] A. Fechtenkötter, K. Saalwächter, M.A. Harbison, K. Müllen, H.-W. Spiess, *Angew. Chem. Int. Ed. Engl.* **38**, 3039 (1999)

[34] K. Weiss, G. Beernink, F. Dötz, A. Birkner, K. Müllen, Ch. Wöll, *Angew. Chem. Int. Ed. Engl.* **38**, 3752 (1999)

[35] U.-M. Wiesler, V. Enkelmann, K. Müllen, *in preparation*

# Tailor-Made Polymer Supports *via* Metathesis Polymerization: Concepts and Applications

Michael R. Buchmeiser*, Frank Sinner, Mathew Mupa

Institute of Analytical Chemistry and Radiochemistry, University of Innsbruck
Innrain 52a, A-6020 Innsbruck, Austria

SUMMARY: Ring opening metathesis polymerization (ROMP) and alkyne polymerization have been applied to the preparation of surface-functionalized inorganic and organic supports using various coating, grafting, precipitation and bulk-polymerization techniques. Well-defined carrier materials in terms of particle diameter, specific surface, pore volume and functionalization have been prepared. The resulting materials have been used for applications in heterogeneous catalysis (e. g. Heck couplings) as well as in bioanalytical (e. g. protein) separations.

## Introduction

The synthesis of supports that possess a well-defined and stable surface chemistry is still an intensively investigated area. Such high definitions are basic requirements for various applications e. g. in heterogeneous catalysis, chromatography, drug delivery, etc. Standard functionalization procedures of organic supports require the use of divergent synthetic approaches that entail a step by step transformation of surface-bound functional groups. In order to avoid such ill-defined syntheses, we elaborated a conceptually new approach based on metathesis polymerization[1] that allows the synthesis of such supports by a convergent synthetic route, starting from functional monomers. In the following, a few recent examples shall be outlined and their general usefulness shall be underlined by a few examples of application.

 CCC 1022-1360/00/$ 17.50+.50/0

**Ring-opening metathesis precipitation polymerization**

Ring-opening metathesis precipitation polymerization may be used to generate beaded, highly cross-linked polymers with a high density of functional groups at the surface. The reaction sequence entails the use of functional monomer. The polymerization system itself must fulfill a class VI living system[2]. Subsequent cross-linking of the "living" linear polymer chains by a suitable norbornene-based cross-linker in course of the precipitation polymerization leads to the formation of irregularly shaped polymer beads (Scheme 1)[3]. Even complex functionalities may be introduced with high reproducibility and without any change in the chemical nature, geometry and even absolute configuration of the corresponding functional group.

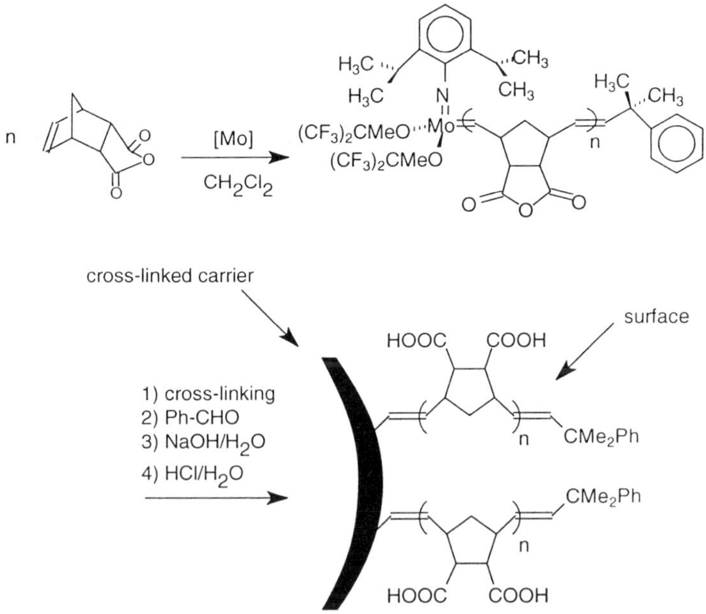


**Scheme 1:** Synthesis of succinic acid derivatized polymer beads by ring-opening metathesis precipitation polymerization[3]. [Mo] = Mo(N-2,6-*i*-$Pr_2$-$C_6H_3$)($CHCMe_2Ph$($OCMe(CF_3)_2)_2$. = cross-linker = 1,4,4a,5,8,8a-hexahydro-1,4,5,8-*exo-endo*-dimethanonaphthalene.

The amount of functional monomer immobilized by this approach may be varied within a range of 0 – 8 mmol/g. Table 1 gives an overview over the functional monomers and beaded materials as well as their physico-chemical properties.

**Table 1.** Poly(norbornene-5,6-dicarboxylic acid)- and poly(norbornene-5-dipyridyl carbamide) ($C(O)Npy_2$)-functionalized resins[3, 4].

| # | functional group | capacity [mmol/g] | swelling[a)] [%] | σ [$m^2/g$] | yield [%] |
|---|---|---|---|---|---|
| **1** | succinic acid | 7.30 | 44 | 30 | 100 |
| **2** | succinic acid | 3.75 | 36 | 14 | 98 |
| **3** | succinic acid | 3.13 | 64 | 24 | 100 |
| **4** | $C(O)Npy_2$ | 1.0 | 39 | 6[b)] /4[c)] | 97.1 |
| **5** | $C(O)Npy_2$ | 0.6 | 35 | 4 | 95.0 |

A-B-block-copolymers (A = functional monomer, B = cross-linker = 1,4,4a,5,8,8a-hexahydro-1,4,5,8-*exo-endo*-dimethanonaphthalene), σ = specific surface, [a)] in water, [b)] from BET measurements, [c)] calculated from GPC data, average particle size = 40±10 μm.

As can be deduced from Table 1, polymer beads prepared by this method only possess a certain microporosity and mainly macropores (> 1000 Å). The particles have a mean diameter of 20 - 40 μm, and are characterized by a low specific surface area ($4 < \sigma < 30\ m^2/g$). Values for σ as determined by BET are in good agreement with those obtained from inverse GPC[5]. Swelling is acceptable (< 40 %) for resins with $\sigma < 10\ m^2/g$. As a consequence of the polymerization order, the linear polymer chains bearing the functional groups form tentacles that are attached to the surface of the cross-linked carrier. This ensures a fast mass transfer with any kind of mobile phase or liquid and lets these materials appear as highly attractive for extraction and immobilization purposes. Thus, poly(norbornene-5,6-dicarboxylic acid)-functionalized resins prepared by ring-opening metathesis precipitation polymerization have been successfully used for the analysis of organic compounds from water[6, 7] and air[8] as well as for the selective enrichment of lanthanides[9]. Addionally, dipyridyl carbamide-based materials may be used as supports for heterogeneous catalysis as it represents a typical small particle catalyst[15]. Thus, the palladium-loaded chelating groups are located exclusively at the surface of the particle[52,60], they are easily accessible, diffusion plays a minor role and coupling reactions proceed within the interphase[65,66]. As a consequence of the high selectivity of the ligand for palladium, no substitution by other metal ions that might be present in technical grade chemicals occurs. The high affinity of the ligand for palladium also leads to a high temperature stability of up to 150°C and to an extraordinary pH-stability ($0 < pH < 10$) of the resulting complex.

**Table 2:** Summary of Heck-couplings and polymerizations[10]. TBAB = tetrabutylammonium bromide. [1)] tri-*n*-butylamine; [2)] $K_2CO_3/Na_2CO_3$ [3)] 0.21 mmol/g immobilized $Pd^{2+}$ (+ 0.4 mmol free ligand), [4)] isolated yields. [5)] 0.35 mmol/g immobilized Pd (0) (+ 0.65 mmol free ligand). [6)] DMAC [7)] DMF. Unless stated otherwise, Heck-couplings were carried out at T = 140 °C, t = 90 h.

| **Ar-X** | **$H_2C{=}CHR$** | **product** | **resin[3)] [mg]** | **yield[4)] [%]** | **TON x $10^3$** |
|---|---|---|---|---|---|
| iodobenzene | styrene | *trans*-stilbene | 1.2[2,6)] | 99.0 | 111.0 |
| iodobenzene | styrene/TBAB | *trans*-stilbene | 1[1,7)] | 98 | 93.2 |
| iodobenzene | ethylacrylate | ethyl cinnamate | 2[1,7)] | 75.0 | 175.1 |
| iodobenzene | vinylferrocene | α-ferrocenylstyrene | 1[1,7)] | 81.1 | 17.4 |
| bromobenzene | styrene | *trans*-stilbene | 3[2,6)] | 90.0 | 40.0 |
| 4-bromobenzonitrile | styrene | 4-cyano-*trans*-stilbene | 1[1,5,7)] | 83.0 | 157.7 |
| 4-bromo-1-fluorobenzene | styrene | 4-fluoro-*trans*-stilbene | 2[2,6)] | 58.5 | 34.4 |
| 4-methoxybromobenzene | styrene | 4-methoxy-*trans*-stilbene | 105[2,6)] | 70.0 | 0.9 |
| chlorobenzene | styrene/TBAB | *trans*-stilbene | 10[2, 5, 6)] | 89.0 | 23.6 |
| 4-chloroacetophenone | styrene/TBAB | 4-acetyl-*trans*-stilbene | 20[2,6)] | 95.0 | 6.1 |
| dibromobenzene | divinylbenzene | PPV | 2[2,6)] | >99 | 18 |

Consequently, this heterogeneous catalyst turned out to be highly active (94 - 99 % yields) in the vinylation of aryl iodides and aryl bromides (Heck couplings) with turn-over numbers (TONs) of up to 210 000[10]. High yields (≤ 95 %) and TONs (≤ 24 000) may additionally be achieved in the tetrabutylammonium bromide (TBAB) assisted vinylation of aryl chlorides (Table 2).

## Ring-opening metathesis graft-copolymerization

Due to the considerable swelling that was observed in ring-opening metathesis precipitation polymerization-based polymer supports with $\sigma > 10\ m^2/g$, we seeked to synthesize entirely pressure-stable ROMP-based supports. For these purposes, a ROMP-based grafting approach for the preparation of surface-functionalized HPLC supports has been elaborated[11]. Bromomethylated, beaded polymer supports based on poly(styrene-divinylbenzene) (PS-DVB) were prepared either *via* direct bromomethylation or *via* transhalogenation of chloromethyl-PS-DVB and subsequently surface-derivatized with norborn-2-ene groups by reaction with sodium norborn-2-ene-5-ylmethanolate. Norborn-2-ene-derivatized silica-based supports were prepared by silanization employing norborn-2-ene-5-yltrichlorosilane. Surface grafting of the norbornene-modified supports with a series of norbornene-based, L-valine and L-phenylalanine-containing chiral monomers was accomplished using ROMP.

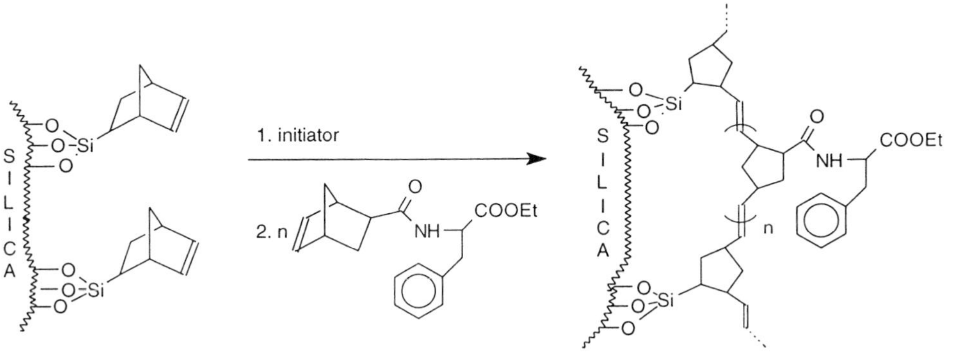


**Figure 2:** Ring-opening metathesis graft copolymerization[11]. Initiator = $Mo(N\text{-}2,6\text{-}Me_2\text{-}C_6H_3)(CHCMe_2Ph(OCMe(CF_3)_2)_2$, $Cl_2(PR_3)_2Ru(=CHPh)$ (R = cyclohexyl).

Both Schrock-type and Grubbs-type initiators were found suitable for that purpose. Scheme 2 gives an overview over the entire reaction sequence. Poly-(N-(norbornene-5-carboxyl)-ß-cyclodextrin ester) as well as poly-(N-(norbornene-5-carboxyl)-phenylalanine ethylester)-surface-grafted, porous 5 μm silica were found suitable for chiral HPLC-separations. In analogy, poly(ethynylferricinium)- and poly(ethynyl-1-(octamethylferrocenylethenyl)benzene)-based anion exchangers have been prepared following the grafting concept described above[12]. Thus, metathesis polymerization of the corresponding terminal alkynes using a Schrock-type catalyst and subsequent grafting of the living polymer onto a NBE-derivatized silica support yielded the desired metallocene-grafted stationary phase. Oxidation of the parent Fe (II) systems to the corresponding ferricinium cations was easily performed in acetonitrile, resulting in a metallocene-based anion-exchanger that has been used successfully for the separation of oligonucleotides.

**Monolithic separation media prepared by ROMP**

Continuous separation media represent highly attractive supports both for separation sciences and heterogeneous catalysis. Since their introduction into separation sciences by Hjertén et al.[13], these materials have been further developed[14]. Organic continuous beds are either based on acrylates or poly(styrene-divinylbenzene) and are almost exclusively prepared by radical polymerization. While the resulting materials have been reported to be well defined in terms of microstructure, specific surface area and pore volume, the preparation of functionalized monoliths is still somehow limited[15]. Due to the high versatility of ring-opening metathesis polymerization (ROMP) for the preparation of functionalized polymer supports we investigated as to which extent transition-metal catalyzed reactions may be used for the synthesis of continuous polymeric supports[47].

ROMP-based monolithic materials have been prepared by ring-opening metathesis copolymerization of norborn-2-ene (NBE) and 1,4,4a,5,8,8a-hexahydro-1,4,5,8-*exo, endo*-dimethanonaphthalene (DMN-H6) within borosilicate columns in the presence of porogenic solvents such as toluene, methylene chloride, methanol and 2-propanol[16] (Figure 3). Grubbs-type initiators of the general formula $Cl_2(PR_3)_2Ru(=CHPh)$ (R = cyclohexyl) were used throughout. The resulting separation media possess microstructures consisting of microporous, spherical microglobules with a narrow size distribution. By variation of the polymerization conditions in terms of stoichiometry of the monomers, porogenic solvents and temperature, microglobule diameters may be varied within a range of 2 ± 1 μm up to 8 ± 2 μm. Specific surface areas ($\sigma$) and inter-microglobule void volumes ($\varepsilon_z$) may be altered within 60 – 190 $m^2/g$ and 5 – 50%, respectively.

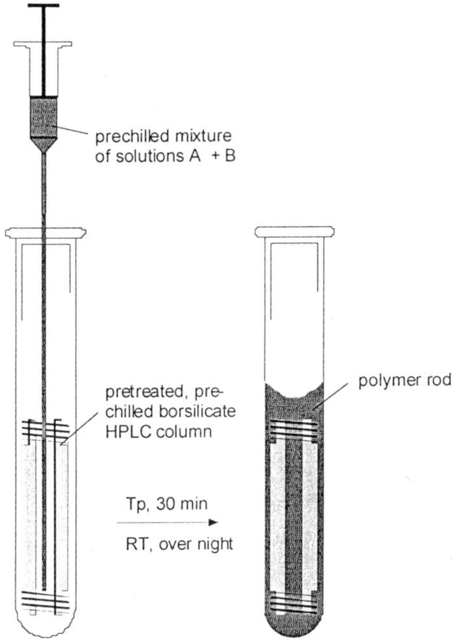


**Figure 3.** Polymerization arrangement for the preparation of molded continuous rods using ROMP.

Table 3 gives an overview over these monoliths. Functionalized continuous rods were synthesized by one additional synthetic step that takes advantage of the living character of the ROMP-based copolymerization. This "in situ" derivatization was achieved after the formation of the continuous rod by reacting the active, surface-bound initiator with functional norborn-2-ene- and 7-oxanorborn-2-ene-based monomers including *endo/exo*-7-oxanorborn-2-ene-5-yl-ß-cyclodextrin, by passing solutions thereof in dichloromethane and DMF, respectively, over the rigid rod (Table 4).

**Table 3.** NBE = norborn-2-ene, DMN-H6= 1,4,4a,5,8,8a-hexahydro-1,4,5,8-*exo*, *endo*-dimethanonaphthalene, **1** = $Cl_2(PR_3)_2Ru(=CHPh)$ (R = cyclohexyl), $T_p$ = polymerization temperature, σ = specific surface, $V_p$ = pore volume[5], $d_p$ = microglobule diameter[5], [1)] by weight.

| # | NBE [%[1)]] | DMN-H6 [%[1)]] | toluene [%[1)]] | 2-PrOH [%[1)]] | 1 [%[1)]] | $T_p$ [°C] | σ [$m^2/g$] | $\varepsilon_z$ [%] | $V_p$ [mL] | $d_p$ [µm] |
|---|---|---|---|---|---|---|---|---|---|---|
| 1 | 20 | 20 | 10 | 50 | 0.4 | 0 | 62 | 33 | 0.31 | 4±1 |
| 2 | 20 | 20 | 20 | 40 | 0.4 | 0 | 74 | 21 | 0.31 | 4±1 |
| 3 | 15 | 15 | 10 | 60 | 0.4 | 0 | 76 | 37 | 0.31 | 2±1 |
| 4 | 25 | 25 | 10 | 40 | 0.4 | 0 | 85 | 15 | 0.34 | 2±1 |
| 5 | 25 | 25 | 10 | 40 | 1 | 0 | 86 | 14 | 0.34 | 4±1 |
| 6 | 25 | 25 | 20 | 30 | 0.4 | 0 | 91 | 15 | 0.33 | 4±1 |
| 7 | 25 | 25 | 10 | 40 | 0.4 | -20 | 98 | 17 | 0.32 | 6±2 |
| 8 | 15 | 15 | 20 | 50 | 0.4 | 0 | 110 | 49 | 0.28 | 3±1 |
| 9 | 30 | 30 | 10 | 30 | 0.4 | 0 | 191 | 5 | 0.35 | 8±2 |

**Table 4.** Capacities of functionalized ROMP monoliths. [a)] determined by titration, [b)] determined by elemental analysis (percent of nitrogen). [1)] by weight.

| # | monomer | capacity [mmol/g] |
|---|---|---|
| 1 | norbornene-2-carboxylic acid | 0.2[a] (3.3 %)[1)] |
| 2 | norbornene-2,3-dicarboxylic acid | 0.14[a] (2.0 %)[1)] |
| 3 | N,N-(norbornene-5,6-dicarboxyyl)valine-N'-(3-nitrophenyl)amide | 0.03[b] (1.1 %)[1)] |
| 4 | N-phenyl-7-oxanorbornene-2,3-dicarbimide | 0.22[b] (5.3 %)[1)] |
| 5 | N-(4-hydroxyphenyl)-7-oxanorbornene-2,3-dicarbimide | 0.06[b] (1.5 %)[1)] |
| 6 | N-(4-dimethylaminophenyl)-7-oxanorbornene-2,3-dicarbimide | 0.26[b] (7.4 %)[1)] |

## Conclusion

It has been shown that ROMP represents a highly versatile tool in the preparation and modification of functional polymer supports. In contrast to standard binary and ternary systems, the use of well-defined initiators allows the highly reproducible preparation of tailor-made materials with regards to particle size, diameter, pore volume and functionalization. Most polymerization techniques such as precipitation- graft- and even bulk polymerization may be used for these purposes.

## References

1. M. R. Buchmeiser, *Chem. Rev.* **in press**, (2000)
2. K. Matyjaszewski, *Macromolecules* **26**, 1787 (1993)
3. M. R. Buchmeiser, N. Atzl, G. K. Bonn, *J. Am. Chem. Soc.* **119**, 9166 (1997)
4. F. Sinner, M. R. Buchmeiser, R. Tessadri, M. Mupa, K. Wurst, G. K. Bonn, *J. Am. Chem. Soc.* **120**, 2790 (1998)
5. I. Halász, K. Martin, *Angew. Chem.* **90**, 954 (1978)
6. D. Ambrose, J. S. Fritz, M. R. Buchmeiser, N. Atzl, G. K. Bonn, *J. Chromatogr. A* **786**, 259 (1997)
7. K. Eder, M. R. Buchmeiser, G. K. Bonn, *J. Chromatogr. A* **810**, 43 (1998)
8. G. Seeber, M. R. Buchmeiser, G. K. Bonn, T. Bertsch, *J. Chromatogr. A* **809**, 121 (1998)
9. M. R. Buchmeiser, R. Tessadri, G. Seeber, G. K. Bonn, *Anal. Chem.* **70**, 2130 (1998)
10. M. R. Buchmeiser, K. Wurst, *J. Am. Chem. Soc.* **121**, 11101 (1999)
11. M. R. Buchmeiser, F. Sinner, M. Mupa, K. Wurst, *Macromolecules* **33**, 32 (2000)
12. K. Eder, H. Schottenberger, C. G. Huber, M. R. Buchmeiser, *unpublished*
13. S. Hjertén, Y.-M. Li, J.-L. Liao, J. Mohammad, K. Nakazato, G. Pettersson, *Nature* **356**, 810 (1992)
14. E. C. Peters, F. Svec, J. M. J. Fréchet, *Adv. Mater.* **11**, 1169 (1999)

15. E. C. Peters, F. Svec, J. M. J. Fréchet, C. Viklund, K. Irgum, *Macromolecules* **32**, 6377 (1999)

16. F. Sinner, M. R. Buchmeiser, *Angew. Chem.* **in press**, (2000)

# Structure Design of Polysaccharides: Novel Concepts, Selective Syntheses, High Value Applications

Dieter Klemm*, Lars Einfeldt

Institute of Organic and Macromolecular Chemistry, University of Jena, Humboldtstraße 10, D-07743 Jena, Germany

SUMMARY: Using protecting silyl groups, catalyzed transesterification, and stereochemically controlled $S_N2$ mechanisms examples of well-defined 2-, 3-, 6-, and 2,6-ethers, esters, resp. amino derivatives of cellulose, starch, and related polysaccharides have been synthesized and characterized by two-dimensional NMR spectroscopy. Supramolecularly structured polymer matrices useful in biosensors could be developed by immobilization of enzymes like glucose oxidase and aromatic redox-chromogenic structures at 6-deoxy-6-(4-aminophenyl)-aminocellulose. Cellulose tubes "grown" biotechnologically from D-glucose in the culture medium of *Acetobacter xylinum* represent novel types of artificial blood vessels because of their unique highly-swollen nanofiber network-structure.

## Introduction

Polysaccharides like cellulose and starch are important renewable biopolymers characterized by high functionality, strong hydrophilicity, remarkable structure-forming potential, chirality, and important biocompatibility.
To use these unique properties efficiently selective functionalization of the hydroxy groups and of subsequently introduced functionalities, high-performance analytical methods, and controlled formation of supramolecular structures are essential.

From this point of view the structure design of polysaccharides means a „tailor-made" modification in the whole scale from Angströms up to centimeters. As demonstrated schematically in Fig. 1 it starts with the regioselective functionalization within the repeating anhydroglucose units and along the polymer chains (molecular structure), leads to well-defined supramolecular films with recognition activity, and includes the formation and application of biotechnologically structured polymeric biomaterials.

 CCC 1022-1360/00/$ 17.50+.50/0

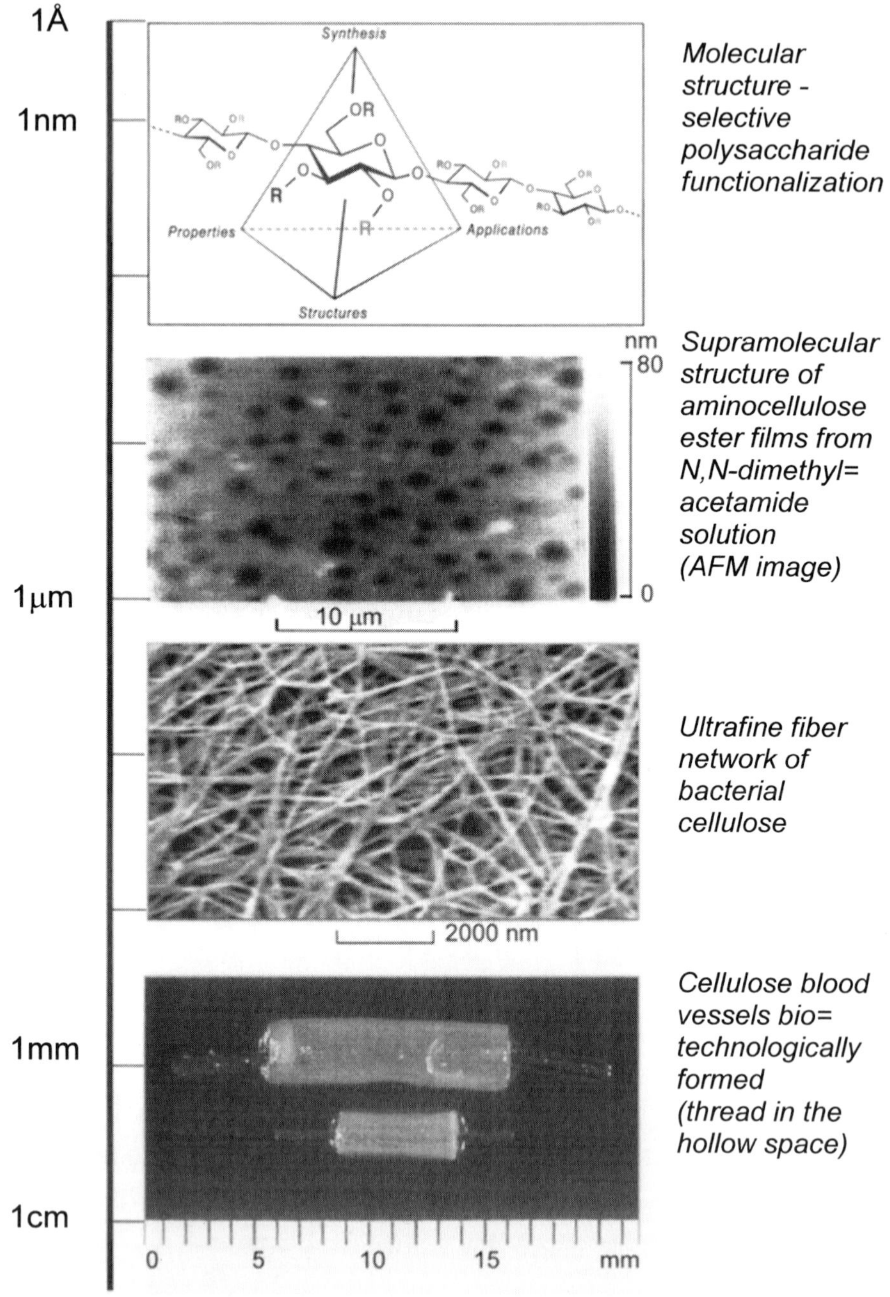


Fig. 1: Structure design of polysaccharides.

## Concepts in regiochemistry of polysaccharides

To realize the site-selective functionalization of the OH-groups of polysaccharides three effective pathways have been developed up to now[1-4]:

- using bulky protecting groups, especially methoxy-substituted triphenylmethyl[5, 6] and different types of silyl groups[7, 8] (e.g., bulky 6- and 2,6-silyl ethers of starch and cellulose),
- selective activation of OH groups by special structures in solution combined with tailored reagents resp. catalysts (e.g., transesterification with carboxylic acid vinylesters to 2-O-esters of polysaccharides)[9-11],
- introducing of functional groups reacting site-selectively according to stereochemically controlled mechanisms ($S_N2$ reactions of cellulose sulfonates with diamines to 6-aminoderivatives)[12, 13].

### *Silyl starches*

As demonstrated in Fig. 2 the homogeneous silylation of starch with bulky thexyldimethylsilylchlorosilane (TDSCl) in dimethylsulfoxide (DMSO)/pyridine leads to a high regioselective 6-O-silylation using equal amounts of the silylating agent.

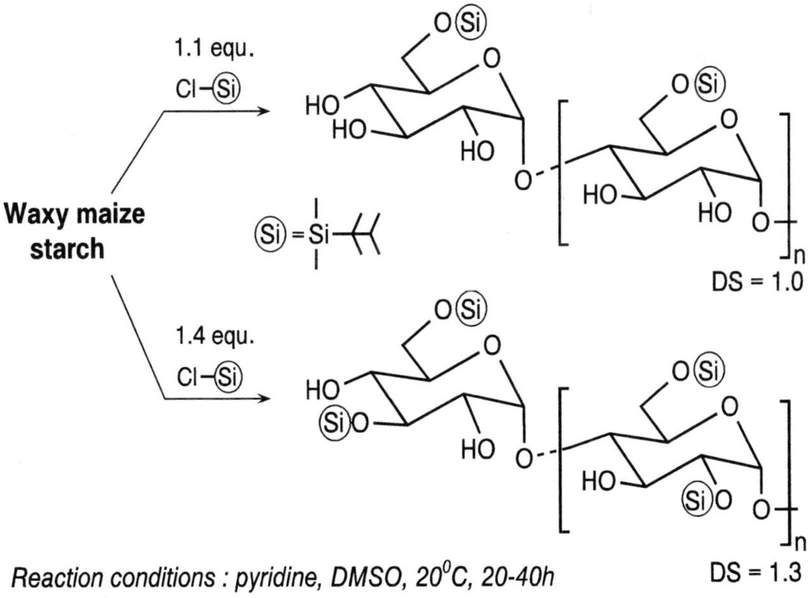


Fig. 2: Synthesis pathways to regioselectively functionalized silyl starches.

With higher equivalents of TDSCl a additional etherification of positions 2 have been obtained as known for the homogeneous silylation of cellulose[7, 8)] and of cyclodextrins, too[14)]. The further reaction of position 2 of the anhydroglucose unit (AGU) was proved exactly by two-dimensional $^1$H-NMR spectroscopy as well as by the reductive cleavage method[14)]. The [$^1$H$^1$H]-cosydqf-NMR spectrum of 2,6-di-O-thexyldimethylsilyl starch (DS 1.3) after complete peracetylation of free OH-groups is represented in Fig. 3.
Due to the branching structure, starch with high amylopectin content possess a high number of non-reducing terminal end groups (TEG). Suprisingly, in comparison to the AGU within the chains the silylation of the TEG showed a regioselective etherification of OH-groups in position 3 after 6-O-silylation[15)].
2,6-di-O-silyl polysaccharides are useful tools for the synthesis of regioselectively substituted alkyl polysaccharides by subsequent etherification and desilylation with tetrabutylammoniumfluoride. Using 2,6-di-O-thexyldimethylsilyl cellulose uniformely substituted 3-O-methyl cellulose, DS = 1.0, has been obtained[16)]. In contrast to these results, the subsequent methylation of the corresponding 2,6-di-O-thexyldimethylsilyl starch leads to the 2-O-methyl ether via migration of the TDS-groups from position 2 to 3[14)]. Different from the AGU within the chains of starch the terminal end groups are methylated additionally in position 3 resp. 4[17)].

The described selective functionalization using protecting groups represents a suitable way to prepare novel types of functionalized polysaccharides with a well-defined molecular structure in a laboratory scale and to make success in basic questions of polysaccharide research. Typical examples of such a kind are detailed spectroscopic characterization, structure in solution, formation of supramolecular structures, properties at interfaces including interactions with enzymes, and structure-property relationships.

From the view-point of industrial important polysaccharide derivatives regioselectively one-pot synthesis pathways are of importance.
With respect to this situation and to avoid protecting groups we have been investigated the selective esterification of polysaccharides controlled by their structure in solution and by selected reagents resp. catalysts.

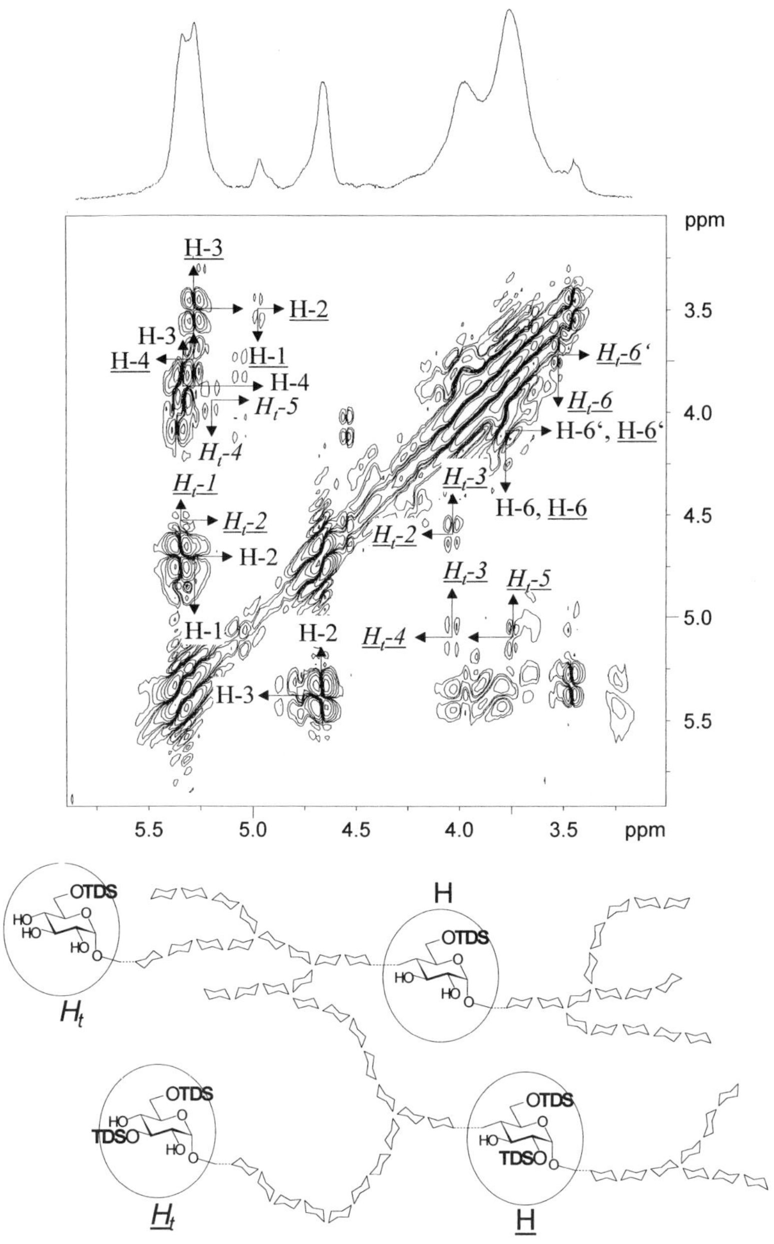


Fig. 3: Structure of peracetylated 2,6-di-O-thexyldimethylsilyl starch, $DS_{Si}$ = 1.3. The cross-peaks of the [$^1$H,$^1$H]-cosydqf-NMR spectrum show the designed silylation of position 6 and than 2 in the AGU in contrast to position 6 and than 3 in the TEG. The arrows in the crosspeaks assign the chemical shifts of the methine and methylene protons of each AGU (H and H) as well as of TEG ($H_t$ and $H_t$).

## *2-O-Esters of polysaccharides*

The regioselective synthesis of 2-O-acyl polysaccharides can be realized by a transesterification method patented for different types of vinylester reagents, catalytic active salts, and polysaccharides[9, 10]. This procedure could be applied on different α-D-glucans like starch, nigeran, pullulan, and cyclodextrin, which are soluble in DMSO. Contrary, β-D-glucans like curdlan or cellulose showed no regioselectivity or low conversions if the polymer was insoluble in DMSO. An exception represents lichenan.

As shown in Fig. 4 aliphatic, aromatic, unsaturated, and additionally functionalized acyl groups can be introduced in position 2 of the polysaccharide molecules with high regioselectivity. Especially the synthesis of acrylates, methacrylates, and cinnemates should be emphasized, because known acylation procedures to unsaturated esters have often led to grafted polymers.

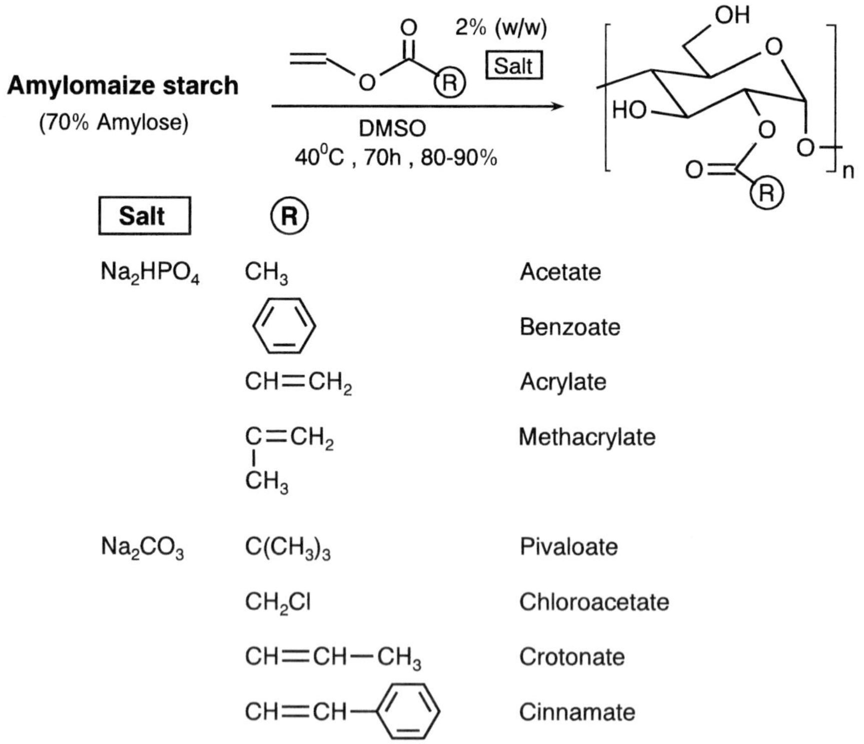


Fig. 4: Regioselective synthesis of 2-O-acyl starches avoiding protecting groups.

A direct proof of the regioselective acylation could be made using two-dimensional NMR spectroscopy (HMBC- and HMQC-techniques) after peracetylation or perpropionylation of the 2-O-acyl polysaccharides. In dependence of the amount of acylating agent a controlled DS in position 2 could be achieved. Such 2-O-acetyl starches with DS = 0.3, DS = 0.8 and DS = 1.0 were used recently for the explaination of specific relaxation processes in the dielectric relaxation spectroscopy[18)].
The pool of catalytic active salts includes further NaCl, $NH_4Cl$, $(NH_4)_2\ CO_3$, $Na_2HPO_4$, $LiCO_3$, and $K_3PO_4$. Without any salt no regioselectivity could be obtained. The detailed mechanism of the catalyzed transesterification is now under investigation.
The regioselectivity is mainly controlled by the structure of polysaccharides in DMSO solution[19)], the vinylester reagent forming acetaldehyde during reaction, and the type of catalytic active salt.

## *6-deoxy-6-aminocelluloses*

An important way to introduce functional groups regioselectively in position 6 of cellulose consists in the nucleophilic substitution reaction of organo-soluble and film-forming p-toluenesulfonic acid esters of cellulose (tosylcelluloses) as intermediates[20, 21)]. A typical example of this pathway is the synthesis of 6-deoxy-6-aminocelluloses using different types of diamines as nucleophilic reagents[12)].
As shown in Fig. 5 tosylcellulose (DS 2.3) reacts with 1,4-phenylenediamine (PDA) under the described conditions to the corresponding 6-deoxy-6-(4-aminophenyl)-amino-2(3)-O-tosylcellulose ("PDA cellulose"). The best results have been obtained in presence of triethylamine with respect to buffering the p-toluenesulfonic acid formed and avoiding of staining the reaction mixture. A typical PDA cellulose with $DS_{PDA}$ 0.75 and $DS_{tosylate}$ 1.30 which was used for subsequent reactions (see next part) was synthesized in presence of this amine at 100 °C within 3 h, employing 9 molar equivalents of PDA per AGU of the tosylcellulose. Under these optimized conditions the monoalkylation of PDA was possible, forming soluble polymers without detectable crosslinking. The substitution reaction of tosylcellulose and PDA follows a $S_N2$ mechanism. Because of the sterical proportions of tosylcellulose and PDA, the formation of the transition complex is obviously possible only at the C-6 position. Thus, the replacement of the tosylate groups by PDA proceeds exclusively at this site[12)].
Based on these results a broad spectrum of PDA cellulose derivatives was synthesized after nearly complete acylation of the free OH groups at tosylcelluloses (DS 0.5-1.9) followed

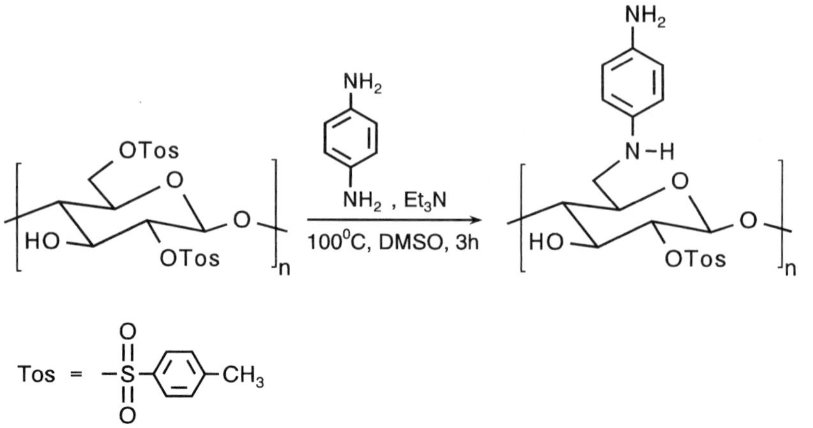


Fig. 5: Reaktion of tosylcellulose (DS 2.3) with 1,4-phenylenediamine.

by the substitution reaction with PDA. Typical ester groups are benzoates and carbanilates. It could be shown that the substitution of the tosylate groups by PDA occured up to more than 90 % after 6 h avoiding a remarkable cleavage of the secondary ester groups[13].

## Supramolecular cellulose architectures with recognition activity

With the aim to develop polymer matrices useful in enzyme sensors the design of polymer supports with the following typical properties was important: solubility in organic solvents, formation of transparent films, anchor groups for immobilization of enzymes and redox-chromogenic structures, enzyme compatibility, and stabilization of redox-chromophors.

As presented schematically in Fig. 6 the aminocelluloses described in the previous part are convenient starting materials to prepare this type of polymers.

All PDA cellulose esters are soluble in DMSO and N,N-dimethylacetamide (DMA) after isolation. During storage, the solubility decreases caused by formation of hydrogen bonds between the amino groups of the PDA unit and residual OH-groups of the cellulose backbone and by crosslinking reactions of the $NH_2$-functions with tosylate residues resp. by autoxidation and subsequent oxidative coupling of the PDA groups[13]. The longest stable derivatives were the PDA cellulose carbanilates (S in Fig. 6: $C_6H_5$-NH-C(O)O).

The PDA cellulose esters form transparent films from the solutions. The supramolecular structure of the films (typical example see Fig. 1) is controlled by the ester groups, the solvent, and the film-forming temperature.

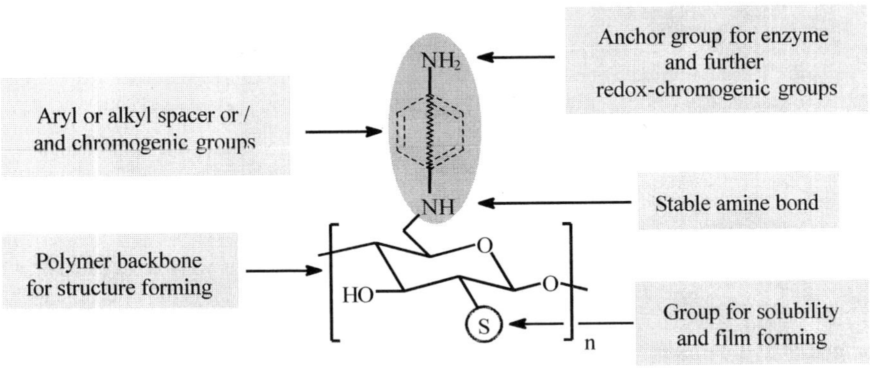


Fig. 6: 6-deoxy-6-Aminocelluloses tailor-made for analyte-sensitive reagent phases.

Onto PDA cellulose derivative films enzymes like glucose oxidase were immobilized[13] using the well-known surface activation by means of glutardialdehyde reaction[22], diazo coupling[23], and the novel ascorbic acid reaction[24]. Fig. 7 shows these and additional procedures used in our investigations.

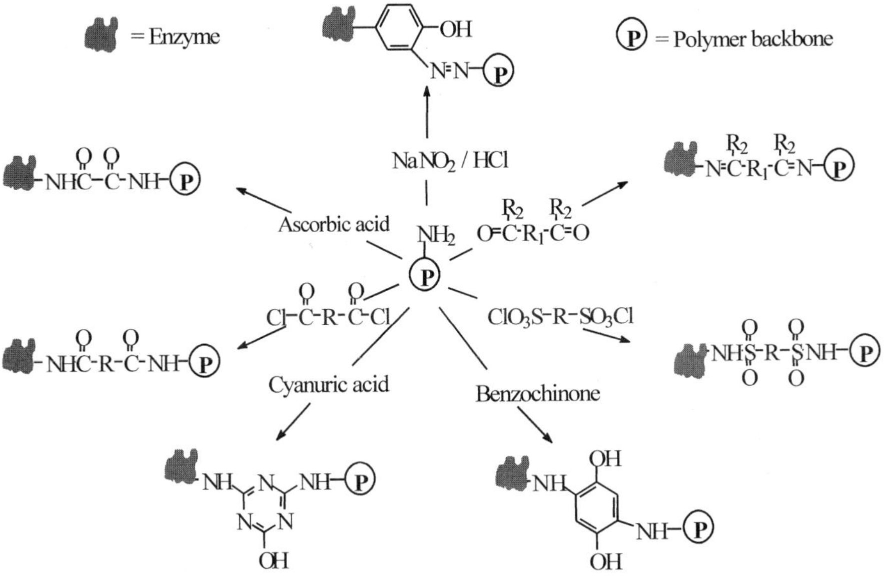


Fig. 7: Covalent enzyme immobilization on aminocelluloses (see Fig. 6) by different types of bifunctional reagents.

In summary, the PDA cellulose esters are suited excellently for enzyme immobilization. The enzymes remain fixed on the films in all cases. No enzyme activity is lost within several days. A typical immobilized enzyme activity is 145 mU/cm$^2$ for coupling with ascorbic acid onto PDA cellulose carbanilate[13].

With respect to the use of PDA cellulose esters as supports for analyte-sensitive reagent phases in fiber-optical biosensors, the redox-chromogenic properties were demonstrated by oxidative coupling reactions of phenols onto the PDA groups in presence of $H_2O_2$ and peroxidase (Fig. 8). The best coloring results were obtained by using PDA cellulose carbanilates.

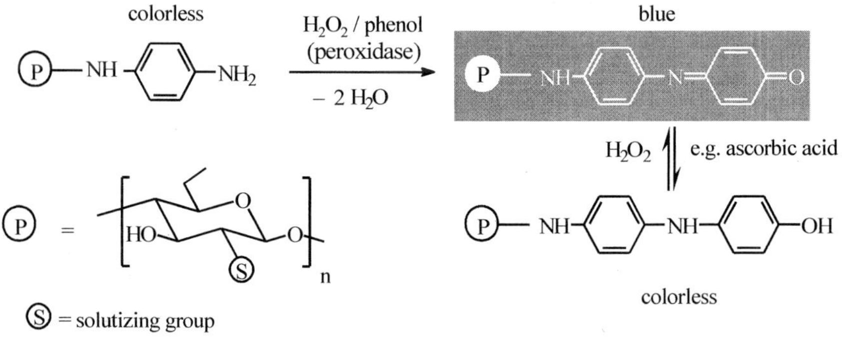


Fig.8: Oxidative coupling of phenol onto PDA cellulose esters (see Fig. 5).

## Biotechnological formation and application of cellulose biomaterials

In addition to plants, bacteria like *Acetobacter xylinum* are able to produce cellulose in the form of pellicles at the air/liquid interface of a static culture medium[25-27]. In contrast to plants, the bacterial cellulose biosynthesis starting from D-glucose offers the very important possibility to control the cellulose formation[27, 28], especially the design of alternative supramolecular structures. Another important advantage of the microbial cellulose is their mouldability in situ.

Based on this biotechnological process we developed the novel biomaterial BASYC-BActerial SYnthesized Cellulose[29, 30].

With a special technology it is possible to prepare miniaturized tubes (hollow fibers) with inner diameters of about 1 mm during the cultivation and without application of physical styling methods (see Fig. 1, bottom). That means, the tubes are "growing" from dextrose directly in the aqueous culture medium. This process is demonstrated schematically in Fig. 9.

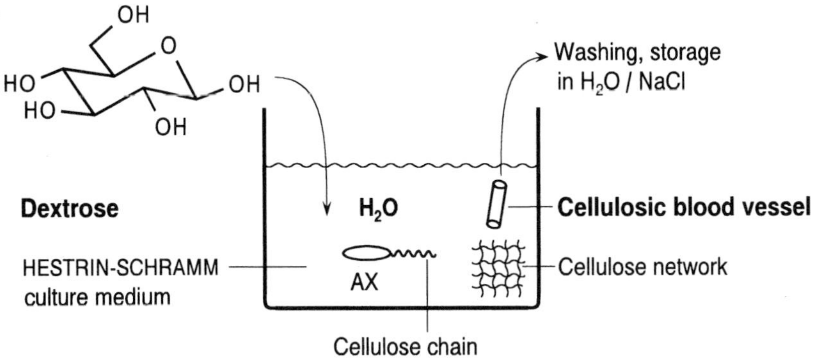


Fig. 9: Schematic representation of direct "growth" of artificial blood vessels in the culture medium of *Acetobacter xylinum.*

The BASYC biomaterial shows an unique supramolecular structure and properties not available starting from plant cellulose or using regeneration processes. The wall of the tubes is formed by a swellable network of nanofibers (see Fig. 1) resulting in a good mechanical strength in the wet state and in a very low roughness of the inner surface.

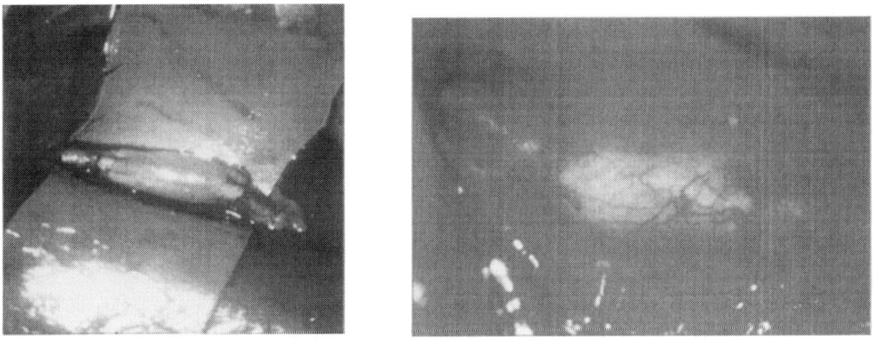

Fig. 10: Carotid artery of the rat with a BASYC interposition immediately after operation (left) and 4 weeks after operation (right).

As demonstrated in animal experiments (rat) the BASYC tubes are useful as "nature-like" blood vessels in microsurgery. The cellulose vessels are biocompatible, not rejected and do not increase the risk of thrombosis. A "vitalization" of the biomaterial (complete overlay of the inside wall with endothelial cells) takes place during a period of about 4 weeks. Connective tissue grows over the outside wall of the cellulose tube integrating it into the body. Fig. 10 shows a BASYC interposition implanted in the carotid artery of the rat[31).

## Conclusion

New concepts in regioselective functionalization, in formation of supramolelcular structures, and in biotechnological preparation enable the "tailor-made" modification of polysaccha- rides in the whole scale from the molecular up to the morphological/ macroscopic level.

Based on these results, new insights in basic questions of polysaccharide research and high value applications, e.g., polymers with recognition activity and novel types of biomaterials are accessible.

## Acknowledgement

We are grateful for the financial support of the Deutsche Forschungsgemeinschaft granted within the Schwerpunktprogramm "Cellulose und Cellulosederivate – molekulares und supramolekulares Strukturdesign".

## References

1. D. Klemm, B. Philipp, Th. Heinze, U. Heinze, W. Wagenknecht, *Comprehensive Cellulose Chemistry*, Wiley-VCH, Weinheim 1998
2. D. Klemm, Th. Heinze, B. Philipp, *Acta Polymerica* **48**, 277 (1997)
3. D. Klemm, W. Wagenknecht, in: *Hydrocolloids-Part 1: Phys. Chem. and Ind. Appl. of Gels, Polysaccharides, and Proteines*, K. Nishinari (Ed.), Elsevier Science, B.V., Amsterdam 2000, p. 251-260
4. B. Philipp, D. Klemm, U. Heinze, *Polymer News* **24**, 305 (1999)
5. M. Yalpani, *Tetrahedron* **41**, 2957 (1985)
6. J.A. Camacho Gomez, U.-W. Erler, D. O. Klemm, *Macromol. Chem. Phys*. **197**, 953 (1996)
7. A. Stein, D. Klemm, *Das Papier* **49**, 732 (1995)
8. A. Koschella, D. Klemm, *Macromol. Symp.* **120**, 115 (1997)
9. R. Dicke, *Ph. D. Theses*, University of Jena 1999
10. DE 19909786.0 (05.03.1999), Wolff Walsrode AG, invs.: E. Klohr, W. Koch, D. Klemm, R. Dicke
11. DE 19951734.7 (27.10.1999), Wolff Walsrode AG, invs.: E. Klohr, W. Koch, D. Klemm, R. Dicke
12. J. Tiller, P. Berlin, D. Klemm, *Macromol. Chem. Phys.* **200,** 1 (1999)
13. J. Tiller, P. Berlin, D. Klemm, *J. Appl. Polym. Sci.* **75**, 904 (2000)
14. P. Mischnick, M. Lange, M. Gohdes, A. Stein, K. Petzold, *Carbohydr. Res*. **277**, 179 (1995)
15. L. Einfeldt, K. Petzold, W. Günther, A. Stein, M. Kussler, D. Klemm, *Macromol. Chem. Phys.*, submitted
16. A. Koschella, *Ph. D. Theses*, University of Jena 2000
17. L. Einfeldt, unpublished results
18. J. Einfeldt, A. Kwasniewski, D. Klemm, R. Dicke, L. Einfeldt, *Polymer* (2000), in press

19. M. St.-Jacques, P.R. Sundararajan, K.J. Taylor, R.H. Marcherssault, *J. Am. Chem. Soc.* **98**, 4386 (1976)
20. K. Rahn, M. Diamantoglou, D. Klemm, H. Berghmans, T. Heinze, *Angew. Macromol. Chem.* **238,** 143 (1996)
21. J. Tiller, *Ph. D. Theses*, University of Jena 1999
22. K. Mosbach, in: *Methods in Enzymology*, Vol. 137, Academic Press, New York 1988, p. 14-111
23. C.G. Beddows, R.A. Mirauer, J.T. Guthrie, *Biotechnol. Bioeng.* **22**, 311 (1980)
24. J. Tiller, P. Berlin, D. Klemm, *Biotechnol. Appl. Biochem.* **30**, 155 (1999)
25. S. Hestrin, *Methods Carbohydr. Chem.* **3**, 4 (1963)
26. P. Ross, R. Mayer, M. Benzimann, *Microbiol. Rev.* **55**, 35 (1991)
27. U. Geyer, Th. Heinze, A. Stein, D. Klemm, S. Marsch, D. Schumann, H.-P. Schmauder, *Int. J. Biol. Macromol.* **16**, 343 (1994)
28. L. Einfeldt, D. Klemm, *J. Carbohydr. Chem.* **16**, 635 (1997)
29. U. Udhardt, *Ph. D. Theses*, University of Jena 2000
30. D. Klemm, U. Udhardt, S. Marsch, D. Schumann, *Polymer News* **24**, 377 (1999)
31. D. Klemm, U. Udhardt, S. Marsch, R. Dicke, D. Schumann. 37$^{th}$ IUPAC Congress, August 14-19, 1999, Berlin

# Silylcellulose from Silylation / Desilylation of Cellullose in Ammonia

Werner Mormann*, Jürgen Demeter+, Thomas Wagner#

Universität-GH Siegen, Fachbereich 8, Laboratorium für Makromolekulare Chemie, Adolf-Reichwein-Str. 2, 57068 Siegen, Germany

+present address BASF AG, 67056 Ludwigshafen

#present address Wolff-Walsrode GmbH, 29655 Walsrode

## Summary

Silylation of cellulose with hexamethyldisilazane (HMDS) in liquid ammonia is described emphasizing mechanistic features. Conditions for complete as well as for stoichiometric partial silylation have been established which proved that the ratio of silylating agent and hydrogen bonding capability of the solvent are crucial for complete silylation. The heterogeneous reaction follows first order kinetics in the presence of saccharine as catalyst indicating that the accessability of hydroxy groups is not the rate determining factor. Stoichiometric desilylation of trimethylsilyl cellulose is also possible in tetrahydrofurane containing ammonia as hydrogen bond donor. Trimethylsilyl celluloses have a solubility window in THF with a DS from 1.6 to 2.6 which is narrower for higher molecular weight of the cellulose. Trimethylsilyl celluloses with DS above 2.7 show thermotropic behaviour.

## Introduction

Conversion of hydroxy to trimethylsiloxy groups is extensively used in organic chemistry as well as in carbohydrate chemistry to impart volatility and to improve solubility in non polar solvents[1]. Silylation of Cellulose has been studied by several research groups using different silylating agents and reaction media. Reactive silylating agents like chlorotrimethylsilane[2-4], N,O-bis(trimethylsilyl)acetamide[5, 6] or hexamethyldisilazane (HMDS)[7] were commonly used.

The lack of solubility and the need of activation of cellulose prior to reaction on one side, the non polar character of trimethylsilyl cellulose on the other require special solvents for the silylation. Normally these reactions are heterogeneous at least in a certain range of conversion. Pyridine[2, 3], dimethylacetamide/LiCl[8], formamide[7], dimethylsulfoxide/paraform-aldehyde[6] or hexamethylphosphoric acid triamide[5, 9] have been used.

Greber and Paschinger were the first to report silylation of cellulose with chlorotrimethyl-

 CCC 1022-1360/00/$ 17.50+.50/0

silane in liquid ammonia. They also reported that silylation with HMDS does not occur under these conditions[4]. Klemm and coworkers used ammonia to activate cellulose in organic solvents like tetrahydrofurane, DMF oder Dimethylsulfoxide and chlorotrimethylsilane[10].

We studied silylation in ammonia in order to combine activation of cellulose and the use of HMDS as silylating agent. HMDS gives ammonia as the only by-product. High boiling solvents or salts as formed in the case of chlorotrimethylsilane decrease stability of silyl cellulose or require tedious purification. Complete silylation of cellulose could be achieved by this procedure for the first time[11].

The present paper describes the synthesis of trimethylsilyl cellulose with predictable and reproducible degrees of silylation from silylation and desilylation with emphasis on mechanistic and reactivity aspects of the reactions involved and deal with some properties of the celluloses obtained.

## Results and discussion

*Activation*

Activation of cellulose with ammonia has been used for a number of reactions. Ammonia replaces the OH---O-hydrogen bonds of cellulose by OH---N-bonds, forming an ammonia-cellulose-complex with one ammonia per anhydroglucose unit (AGU) being intercalated [12]. Liquid ammonia treated (LAT) cellulose has increased reactivity as reported by Schleicher et al. in a study of heterogeneous reactions of LAT cellulose [13].

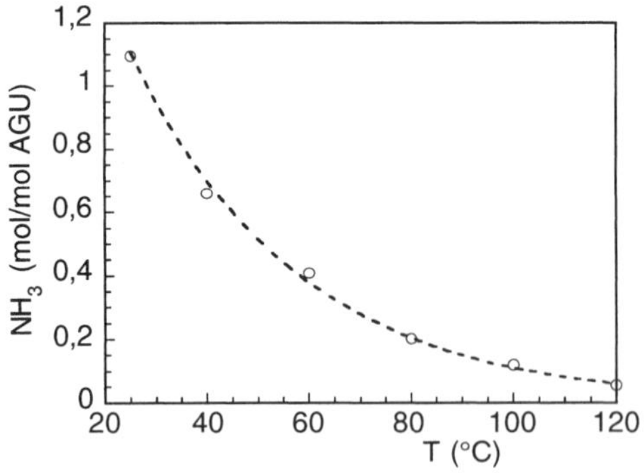


**Fig. 1:** $NH_3$-uptake of linters with temperature

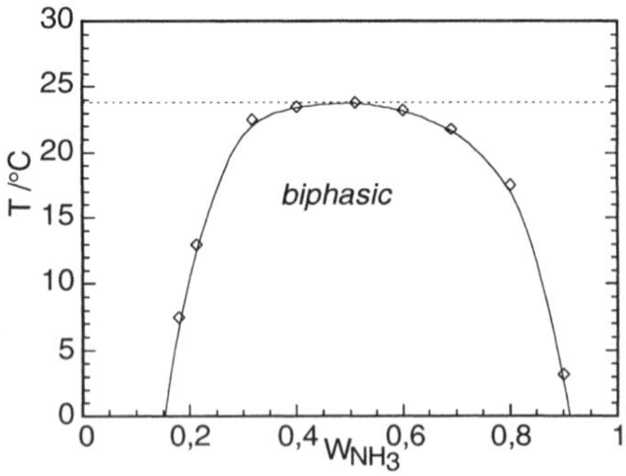


**Fig.2:** Phase diagramme of HMDS/$NH_3$

Saturation of dried cellulose (Linters, Avicel) with gaseous ammonia or with liquid ammonia in an argon atmosphere results in an ammonia/AGU ratio of approximately 1 : 1. With increasing temperature the ammonia content continuously decreases to a ratio of only 1 : 10 at 120 °C as shown in Fig. 1. Obviously the cellulose-ammonia-complex is a temperature dependend equilibrium rather than a stoichiometric compound.

Miscibility studies of ammonia and HMDS revealed complete miscibility above 24 °C and two phase behaviour below this temperature (Fig. 2)[14]. This also explains why Greber [4] was not able to achieve silylation with HMDS in liquid ammonia.

*Silylation with HMDS/ammonia*

Details of the silylation of cellulose as shown in equation 1 and of other hydroxy group containing polymers and low molar mass compounds have been reported previously [11, 14].

Eq. 1:

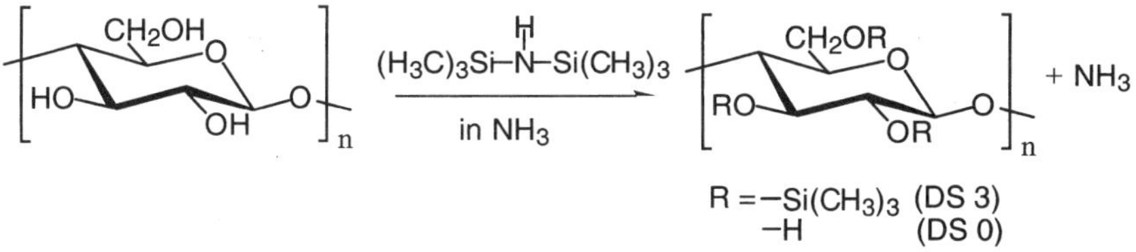


Complete silylation could not be achieved with high excess of HMDS, long reaction times or high temperature. The maximum degree of silylation proved to be independent also on the crystallinity of the cellulose used[15]. Use of saccharin (2-sulfo-benzoic acid imide) as catalyst[16] also had no effect. Surprisingly, use of saccharin and variation of the ammonia / HMDS ratio led to conditions for complete silylation (Tab. 1). Decrease of the amount of ammonia from 20 to 2 mol per mol OH (to 0.4 in the case of Avicel) with constant amount of HMDS (1.1 mol per mol OH) gave complete silylation. A higher excess of HMDS on the other hand leads to a decrease of the degree of silylation which rules out that an equilibrium is responsible for incomplete silylation with higher excess of ammonia.

Trimethylsilyl cellulose with a DS of 2.5 is soluble in HMDS. Attempts were made therefore to obtain fully silylated cellulose by further silylation of such a cellulose. The results are presented in Fig. 3. After 4 days at 80 °C the DS has increased from 2.6 to 2.75 without further reaction after 15 days. Addition of triethylamine as hydrogen bond acceptor enabled silylation up to a DS of 2.95 after additional 20 days with no change after 50 more days. It is worth mentioning that as the DS increased beyond 2.75 the reaction mixture became opalescent and turned into a suspension of trimethylsilyl cellulose in HMDS / triethylamine at even higher conversion.

**Tab. 1:** Complete silylation of cellulose with HMDS/ammonia

| No. | Cellulose | $n_{OH}$ | : | $n_{Si}$ | : | $n_{NH_3}$ | DS |
|---|---|---|---|---|---|---|---|
| 1 | Avicel PH 101 | 1 | : | 3.4 | : | 30 | 2.72 |
| 2 | " | 1 | : | 3.4 | : | 17 | 2.87 |
| 3 | " | 1 | : | 3.4 | : | 9 | 2.94 |
| 4 | " | 1 | : | 3.4 | : | 2 | 3.0 |
| 5 | " | 1 | : | 3.4 | : | 0.4 | 3.0 |
| 6 | " | 1 | : | 13.5 | : | 2 | 2.96 |
| 7 | Avicel PH 102 | 1 | : | 3.4 | : | 2 | 3.0 |
| 8 | Linters | 1 | : | 3.4 | : | 2 | 3.0 |
| 9 | Linters DS 2.57 | 1 | : | 78 | : | 24.3 | 2.95 |
| 10 | Linters DS 2.95 | 1 | : | 78 | : | 60 | 3.0 |

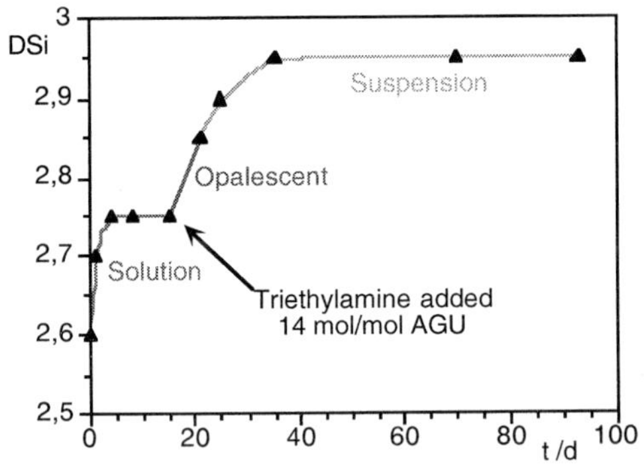


**Fig. 3:** Effect of hydrogen bonds on silylation of cellulose

"Post-silylation" of silyl celluloses was performed with HMDS in ammonia (Tab. 1, entries 9 and 10). It resulted in complete silylation when the ratio of HMDS to ammonia was 2.6 : 1 while the DS was 2.95 for a ratio of 6.4 : 1. Obviously it is the ratio of HMDS / ammonia which is important and not the absolute amounts of them as long as there is enough silylation agent present.

Once reaction conditions for complete silylation were established, the kinetics of this heterogeneous reaction at 80 °C was studied using saccharine as catalyst, HMDS sufficient for a DS of 1.68, and 60 mol ammonia per anhydroglucose unit (AGU). The results are given in Tab. 2 and in Fig. 4. For each data point a silylation was made. A THF soluble and an insoluble fraction was obtained in all cases. The insoluble fractions had a DS of 0.7 and less while those of the soluble fractions increased from 1.7 to 2.3.

Tab. 2: Kinetics of silylation of Avicel with HMDS/ammonia

| t / min | THF-sol. / w.-% | DS THF-sol. | DS insol. | DS total |
|---|---|---|---|---|
| 30 | 41.6 | 1.70 | 0.19 | 0.65 |
| 60 | 66.8 | 1.73 | 0.71 | 1.32 |
| 120 | 74.0 | 2.22 | 0.41 | 1.54 |
| 180 | 69.0 | 2.31 | 0.57 | 1.57 |
| 360 | 72.4 | 2.33 | 0.42 | 1.58 |

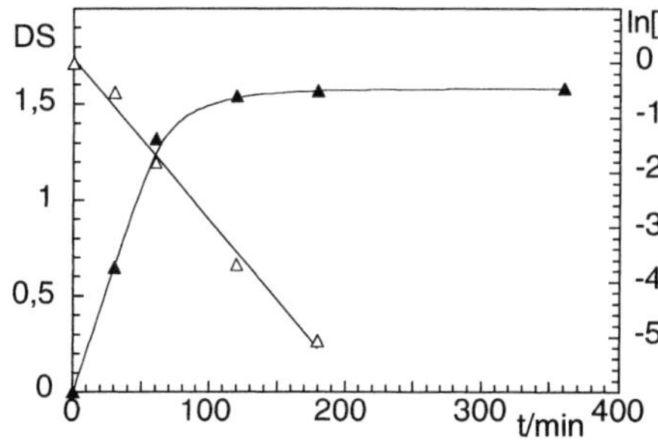


**Fig. 4:** Kinetics of silylation of Avicel

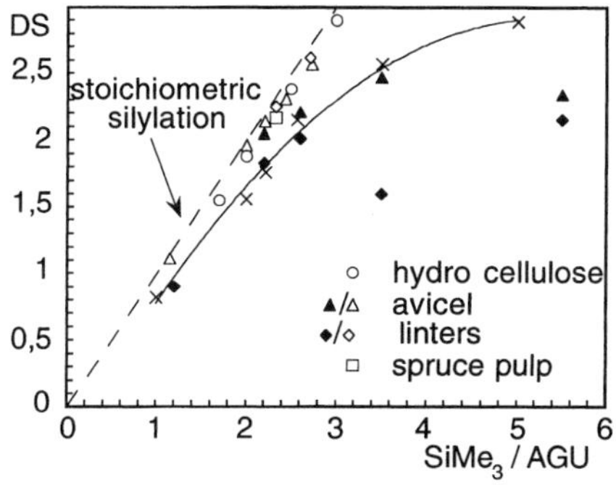


**Fig. 5:** Partial silylation of cellulose (open HMDS, black chlorotrimethylsilane)

Kinetic analysis showed that silylation of cellulose in ammonia under these conditions is of first order for the formation of trimethylsiloxy groups in cellulose. First order kinetics have been found also for the silylation of poly(vinyl alcohol) under similar conditions[17] which, however is a homogeneous reaction. Obviously the active silylating agent is trimethylsilylated saccharin (cf. scheme 1) which is formed in a fast reaction to ensure constant concentration. First order behaviour for the hydroxy groups of cellulose requires that accessability of hydroxy groups is high enough to have no influence on the rate of reaction up to a DS of 1.6. Furthermore 3 hours are a sufficiently long reaction time for complete reaction of all HMDS present in the mixture.

As HMDS reacted completely in the silylation, partial silylation of cellulose was attempted (cf. Fig. 5). In the range between 1.1 and 2.9 degrees of substitution can predictibly obtained with stoichiometric amounts of HMDS[18]. Use of chlorotrimethylsilane under conditions reported in the literature requires excess silylation agent[19].

*Desilylation of trimethylsilyl cellulose*

Desilylation is an important process for regeneration of cellulose. It is normally done with alcohols in the presence of acidic or basic catalysts. Attempts to achieve partial desilylation to a lower DS in THF-solutions were not successful. Again ammonia as hydrogen bond acceptor and donor enabled stoichiometric desilylation in 10 % solutions of silyl cellulose in THF containing 10 % ammonia and saccharin as catalyst. DS below 1 cannot be obtained by this method probably because trimethylsilyl cellulose with a DS of 1 becomes insoluble in the reaction mixture.

**Tab. 4:** Stoichiometric desilylation of trimethylsilyl cellulose (Avicel)

| Run | DS start | DS calc. | DS observed |
|---|---|---|---|
| 1 | 2.96 | 2.3 | 2.29 |
| 2 | 2.79 | 2.0 | 1.96 |
| 3 | 2.79 | 1.9 | 1.93 |
| 4 | 2.96 | 1.7 | 1.63 |
| 5* | 2.96 | 2.0 | 2.43 |
| 6 | 2.92 | 1.0 | 1.13 |
| 7 | 2.92 | 0.7 | 0.96 |
| 8 | 2.96 | 0.3 | 1.06 |

*: without ammonia

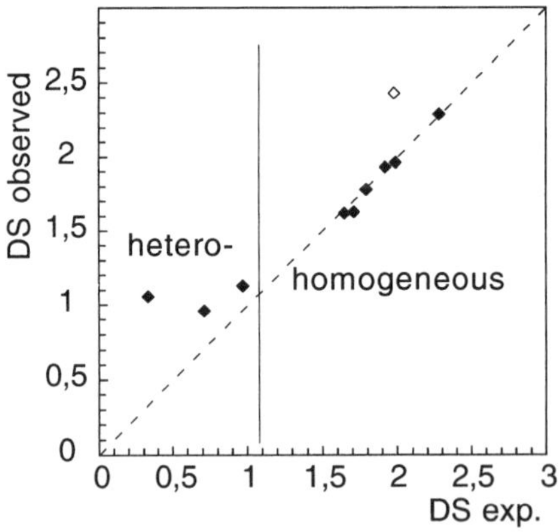


Saccharin which has been used first by Bruynes and Jurriens as catalyst for the silylation with HMDS is also a very effective catalyst in desilylation[16]. The active species most likely is silylated saccharin. We think the combined action as silyl group donor and proton acceptor is essential in silylation while it is silyl group acceptor and proton donor in the desilylation. The rather long reaction times are caused by the slow combination of trimethylsilanol to hexamethyldisiloxane and water.

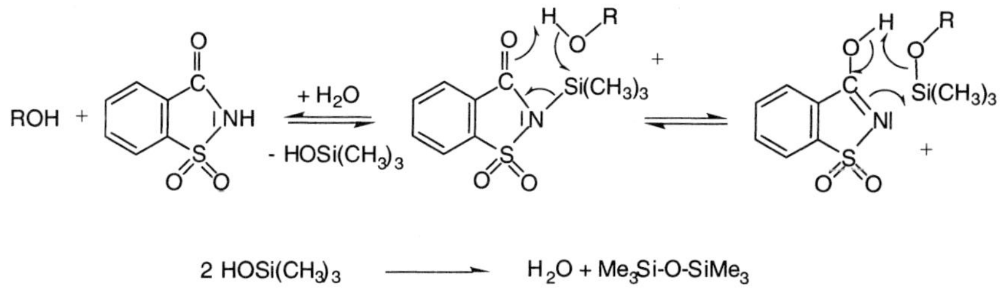


*Properties of trimethylsilyl celluloses*

Silylation as well as desilylation are polymer analogous reactions which has been shown by SEC analysis of suitable derivatives. Thus cotton linters has been converted into the tricarbanilate by reaction with phenyl isocyanate while another sample was silylated, desilylated and then also transformed into the tricarbanilate [14].

Trimethylsilyl celluloses from the silylation reaction described have some peculiarities in the solubility (Fig. 6). They become partly insoluble in THF with a DS below 1,8 and also with high degrees of substitution. The window of solubility becomes smaller with increasing molar mass of the cellulose. Avicel with a DS of 3.0 is soluble only to about 4 % in THF.

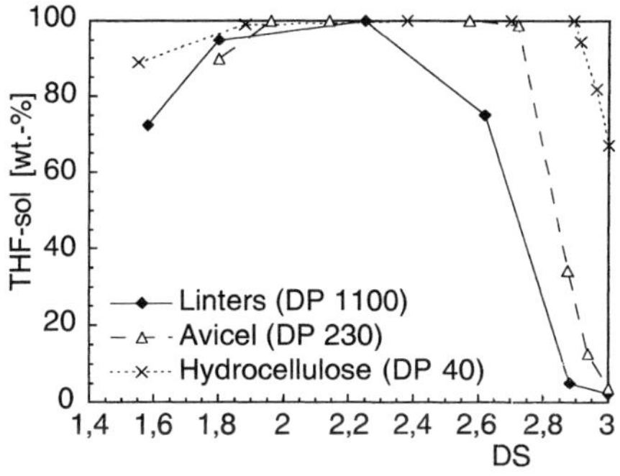


**Fig. 6:** Solubility vs DS of different silyl celluloses

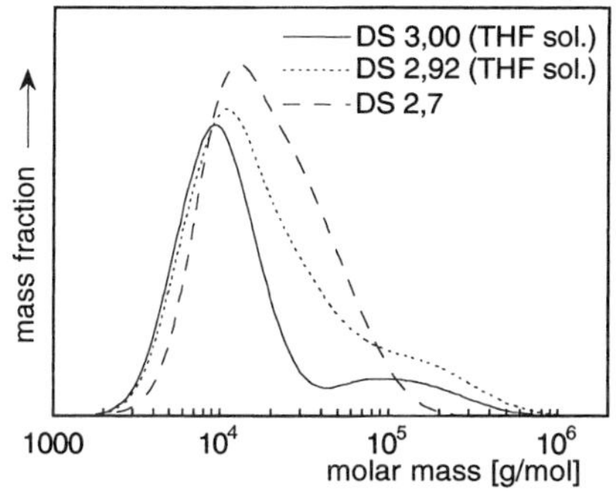


**Fig.7:** Molar mass distribution of Avicel with different degrees of silylation

SEC proved that only the low molar mass fraction is soluble in THF (Fig. 7). The THF-soluble fraction of Avicel (DS 3.0) has a much lower molar mass than a completely soluble non fractionated sample of silylated Avicel. The poor solubility of highly substituted trimethylsilyl cellulose is due to aggregation via interaction of the silyl groups.

Aggregation shows from the bimodal mass distribution curves of highly silylated cellulose. Branching of cellulose molecules via covalent bonds can be excluded as carbanilates obtained after complete desilylation and subsequent reaction with phenyl isocyanate have a monomodal distribution. Cellulosetricarbanilates are known to be molecularly dispersed in dilute solution[20].

Thermal properties of trimethylsilyl celluloses also depend from the degree of substitution. Soluble silyl celluloses behave like normal thermoplastic materials in dynamic mechanic thermal analysis (Fig. 8). They show a glass transition between 190 und 210 °C and a rubber plateau for linters which has five times the molar mass of Avicel. Linters and other highly silylated celluloses show only a slight decrease in modulus above the glass transition. This is in agreement with the results from differential scanning calorimetry.

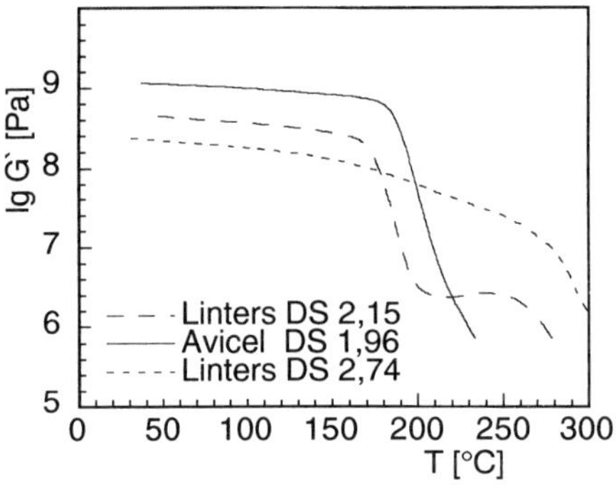


**Fig. 8:** DMTA of silyl celluloses with different DS

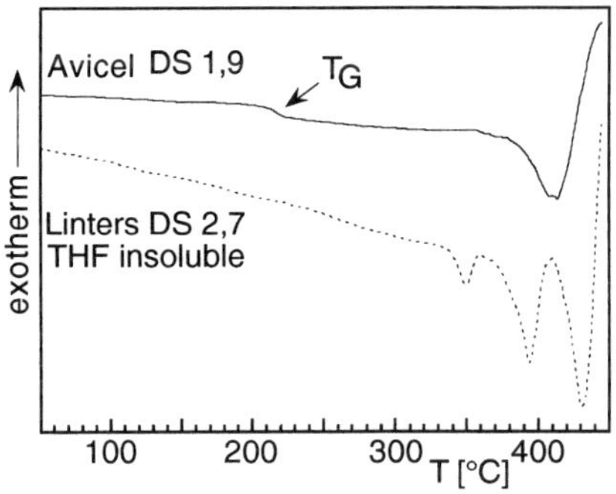


**Fig. 9:** DSC of silyl celluloses with different DS

The DSC-traces (Fig. 9) have a glass transition at about 200 °C and decomposition at about 400 °C for trimethylsilyl cellulose with a DS below 2.5. The insoluble trimethylsilyl cellulose does not show a glass transition but an endotherm which can still be seen after multiple heating/cooling. Up to 360 °C no significant weight loss can be detected from thermogravimetric analysis. Enthalpies of 1-6 kJ/mol as well as the high viscosity above TG suggest that the endotherm is caused by a liquid crystalline transition.

## Conclusions

Silylation of cellulose with HMDS / ammonia is a new variation which allowed the synthesis of completely silylated trimethylsilyl cellulose. Partially silylated cellulose could be obtained in a reproducible manner by stoichiometric silylation as well as by stoichiometric desilylation.

## Acknowledgement

Financial support of this work by the DFG (Forschungsschwerpunkt Cellulose) and the Fonds der Chemischen Industrie is gratefully acknowledged.

## References

1) E. W. Colvin, "Silicon Reagents in Organic Synthesis", in: Best Synthetic Methods, 1st edition, A.R. Katritzky, O. Cohn, C. W. Rees, Eds., Academic Press, New York 1988

2) H.A. Schuyten, J.W. Weaver, J.D. Reid, J.F. Jurgen,. *J. Am. Chem. Soc.* **70**, 1919 (1948)

3) G. Keilich, K. Tihlarik, E. Husemann, *Makromol. Chem.* **120**, 87 (1968)

4) G. Greber, O. Paschinger, *Das Papier* **35**, 547 (1981)

5) J.F. Klebe, H.L. Finkbeiner, *J. Polym. Sci., Part A1* **7**, 1947 (1969)

6) N. Shiraishi, Y. Miyagi, S. Yamashita, T. Yokota, Y. Hayashi, *Sen'i Gakkaishi* **35**, 466 (1979)

7) R.E. Harmon, K.K. De, S.K. Gupta, *Staerke* **25**, 429 (1973)

8) W. Schempp, T. Krause, U. Seifried, A. Koura, *Das Papier* **38**, 607 (1984)

9) J.F. Klebe, H. Finkbeiner, D.M. White, *J. Am. Chem. Soc.* **88**, 3390 (1966)

10) D. Klemm, M. Schnabelrauch, A. Stein, T. Heinze, U. Erler, S. Vogt., *Das Papier* **45**, 773 (1991)

11) W. Mormann, J. Demeter, *Macromolecules* **32**, 1706 -1710 (1999)

12) K. Hess, J. Gundermann, *Ber. Dt. Chem. Ges.* **70**, 1788 (1937)

13) H. Schleicher, C. Daniels, B. Philipp.,. *J. Polym. Sci. Symp.* **47**, 251-260 (1974)

14) W. Mormann, T. Wagner., *Makromol. Chem., Rapid Commun.* **18**, 515 (1997)

15) W. Mormann, T. Wagner., *Carbohydrate polymers* **42**, in press

16) C.A. Bruynes, T.K. Jurriens, *J. Org. Chem.* **47**, 3966 (1982)

17) W. Mormann, T. Wagner, *J. Polym. Sci., Part A: Polym. Chem.* **33**, 1119 (1995)

18) W. Mormann, J. Demeter, T. Wagner, *Macromol. Chem. and Phys.* **200**, 693 (1999)

19) Ger(East)DD 298,644 (1989), invs.: A. Stein, W. Wagenknecht, D. Klemm, B. Philipp, *Chem. Abstr.* **107**, P 92503 b

20) L. Schulz, W. Burchard, *Das Papier* **47**, 1 (1993)

# Polymeric Surfactants from Styrene-co-Maleic-Anhydride Copolymers

Dietrich Braun*, Rainer Sauerwein, Goetz P. Hellmann
Deutsches Kunststoff-Institut, D-64289 Darmstadt, Germany

**Abstract:**

Polymeric surfactants are usually prepared via copolymerisation of suitably substituted monomers. In this study, polymer surfactants were made by modifying a highly functionalized polymer chain back bone: a copolymer of styrene and maleic anhydride, SMA (34wt.-% MA), was turned into anionic, cationic and nonionic surfactants by saponification and imidation of the MA functions. In some cases, the chains were partly hydrophobized by stearyl substituents. All surfactants were water soluble without forming hydrogels. To different degrees, they proved themselves as dispersants and thickeners. Carbon black and titanium dioxide were excellently dispersed by the ionic derivatives o SMA while latices were particularly well thickened by hydrophobized nonionic derivatives.

## 1 Introduction

Polymeric surfactants[1] are water soluble but sufficiently amphiphilic to be adsorbed by surfaces and interfaces, in particular by dispersed solid or liquid phases. This balance is indicated in Fig.1: the polymer chains either cling to the dispersed particles or float between them in the water.

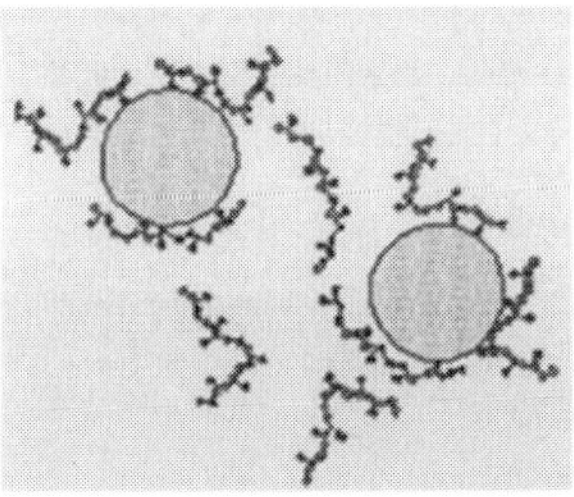

*Fig.1.*
*Aqueous dispersion: phase domains and chains of a polymeric surfactant on them and in the water*

Two important roles of polymeric surfactants are illustrated by Fig.2: *dispersant*[2] (Fig.2a) break up agglomerates of pigments, covering the single particles with a dense adsorption layer; *thickeners*[3,4] (Fig.2b) increase the viscosity of latices, forming a physical chain network in the water that bridges the particles indirectly. A third role is missing in the figure: *flocculants*[5]

 CCC 1022-1360/00/$ 17.50+.50/0

agglomerate and precipitate dispersed particles (Fig.2a reversed), by forming direct chain bridges between the particles. Dispersants and thickeners are discussed in this paper, but not flocculants which need higher molecular weights.

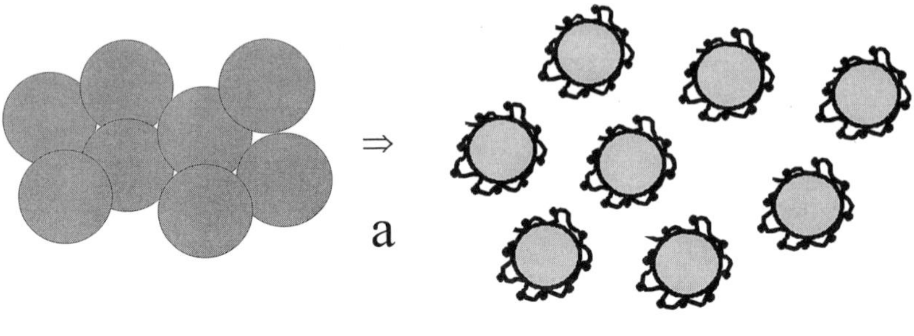


*Fig.2.*
*Roles of polymeric surfactants:*

*(a) dispersing pigments,*
*(b) thickening latices*

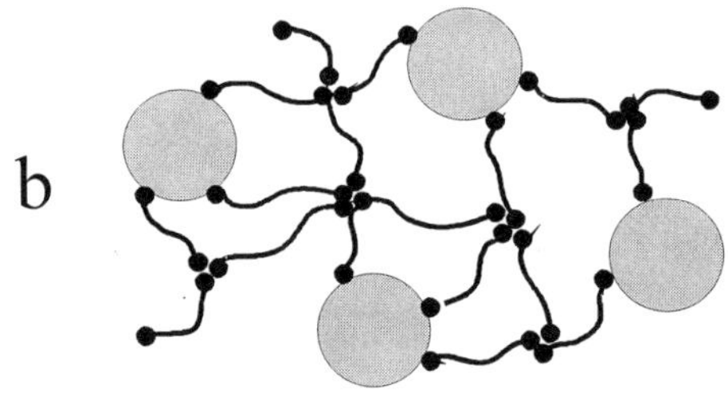


To acquire sufficient water solubility, the chains of polymeric surfactants must inherently be very polar or must be equipped with ionic charges or polar substituents in high concentration. Conventionally, these chains are synthesized by copolymerizing the monomers that are substituted accordingly. But in this study, polymeric surfactants were prepared by polymer analogous chemistry: a maleinated polystyrene, SMA (DSM, The Netherlands), with 34wt.-% of maleic anhydride copolymerized randomly into the styrene chains, was used as a back bone chain to which water soluble substituents were attached (Fig.3).

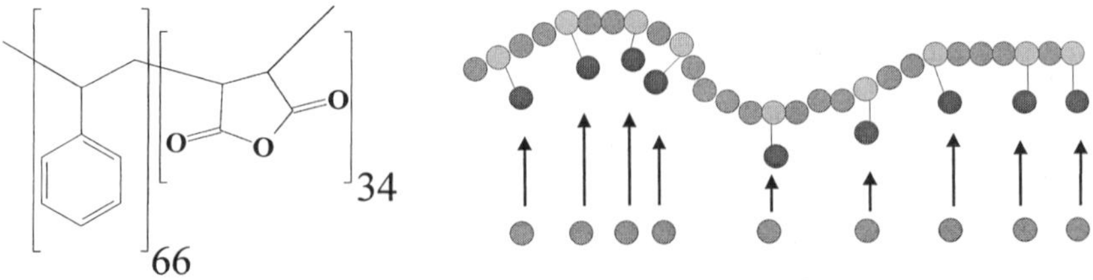

*Fig.3. Maleinated polystyrene SMA; surfactant derivatives were made by adding polar substituents to the anhydride functions along the chains*

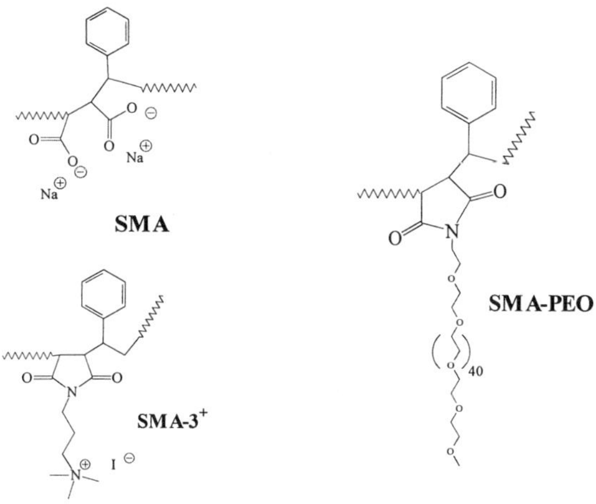


*Fig.4. Derivatives of SMA: anionic $SMA^-$ with carboxylate, cationic $SMA\text{-}3^+$ with ammonium and nonionic SMA-PEO with oligoethyleneoxide groups*

The polymer SMA (molecular weight $M_w=110\times10^3$) is a thermoplastic that is used, often in homogeneous blends with SAN, because of its high glass transition temperature of $T_g=178°C$. But with every third monomer unit carrying an anhydride function, SMA is evidently an attractive polyfunctional back bone for modified products as well.

## 2 Surfactants from a Copolymer of Styrene and Maleic Anhydride

Surfactants derived from SMA are shown in Fig.4: anionic chains ($SMA^-$) were made by saponification of the anhydride functions and cationic ($SMA\text{-}3^+$) and nonionic (SMA-PEO) chains by imidation with an ammonium salt or amino-terminated oligoethyleneoxide. These surfactants are amphiphilic because, while the special substituents (Fig.4) are definitely hydrophilic, the styrene unizs in the back bone are hydrophobic. To strengthen the hydrophobic character, $SMA^-$, $SMA\text{-}3^+$ and SMA-PEO chains were sometimes hydrophobized by incorporating, randomly, stearyl substituents that were added to the MSA chains by imidisation with stearylamin. Particularly interesting were the derivatives of SMA-PEO, i.e. $SMA\text{-}PEO_x$ (x/(1-x): molar ratio of stearyl and PEO units).

The polymeric surfactants thus prepared were soluble in water, while the hydrophobized derivatives exhibited decreasing solubility. None of these surfactants formed hydrogels in water: as shown in Fig.5, the viscosity of water is only slightly increased by $SMA^-$ or SMA-PEO.

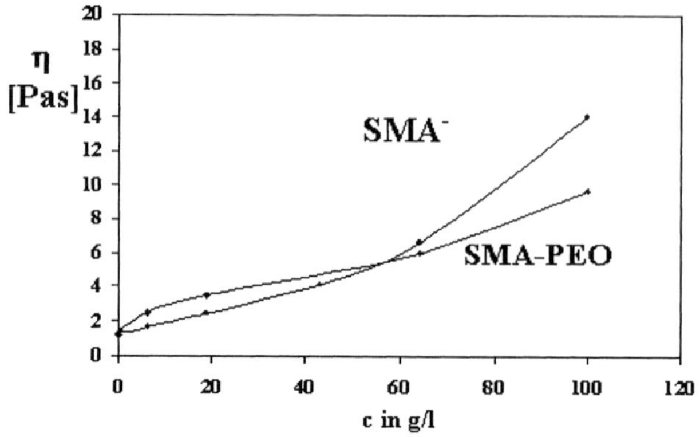


*Fig.5. Viscosity of aqueous surfactant solutions as a function of the concentration c: no hydrogel effect*

## 2.1 Dispersing of Pigments

To test the dispersive power of the surfactants, two pigments were chosen: a graphitized furnace carbon black (Printex L, Degussa, particle diameter 23nm) with a slightly negatively charged surface (charge density -3.85mmol/g) and a strongly negatively charged $TiO_2$ (Bayertitan R-KB-3, Bayer, particle diameter 200-250nm, charge density -12.37mmol/g).

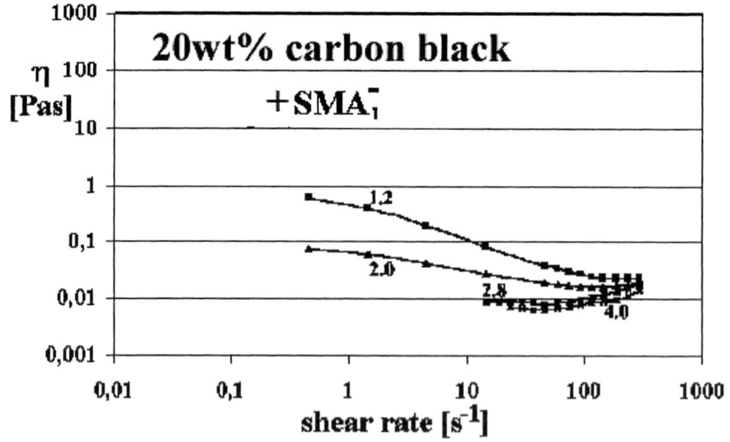


*Fig.6. Viscosity of an aqueous carbon black (20wt.-%) dispersion with slightly hydrophobized $SMA_1^-$ (1mol-% stearyl units); curve parameter: surfactant concentration in wt.-%*

In practice, less carbon black is used than $TiO_2$, in aqueous dispersions. Dispersions with typical concentrations were investigated, i.e. 20wt.-% carbon black and 60wt.-% $TiO_2$.

In aqueous dispersion, the size of the primary particles of these pigments is not relevant because of extensive agglomeration. The carbon black formed in water untractable dispersions with the highly viscous consistency of shoe shine. But small amounts of $SMA^-$ produced already very fluid dispersions, as shown in Fig.6. Slight hydrophobisation proved useful. With 4wt.-% of $SMA^-_1$, the viscosity was not far from that of pure water. The other surfactants, $SMA\text{-}3^+$ and SMA-PEO, were less efficient.

1.2wt.-%

50µm

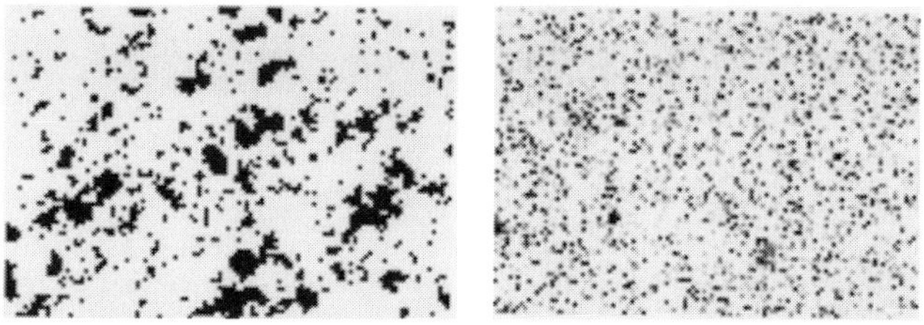

2.8wt.-%

*Fig.7. Carbon black aggregates at different concentrations of $SMA^-$ (light microscopy)*

As demonstrated by Fig.7, the carbon black is well dispersed by $SMA^-$.

Similarly satisfactory results were obtained with $TiO_2$ dispersions, however, with the cationic surfactant $SMA\text{-}3^+$. Figure 8 proves that very small amounts of the surfactant were sufficient to bring the viscosity down to fluid, almost Newtonian behaviour. That the cationic $SMA\text{-}3^+$ was so efficient reflects the fact that the $TiO_2$ carried had a strong negative surface charge.

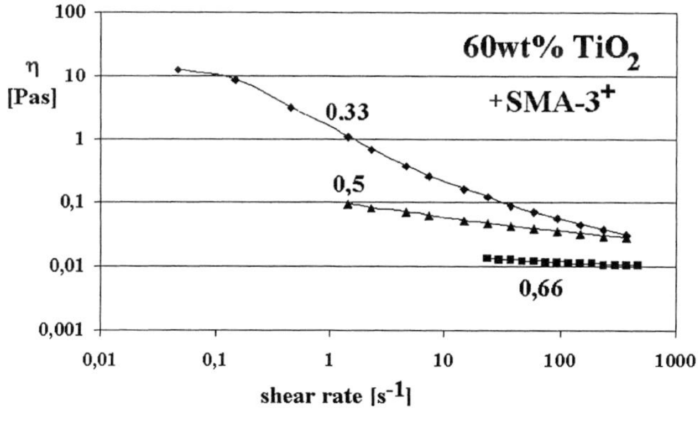


*Fig.8. Viscosity of an aqueous $TiO_2$ (60wt.-%) dispersion with $SMA\text{-}3^+$; curve parameter: surfactant concentration in wt.-%*

## 2.2 Thickening of Latices

To study latex thickening, a PMMA latex was prepared using SDS (sodium dodecylsulfat) as an emulgator so the particles (180nm diameter) were negatively charged. As shown in Fig.9, this latex (35wt.-% polymer) was very fluid, exhibiting Newtonian behaviour.

The nonionic surfactant SMA-PEO proved to be particularly efficient in thickening this latex.

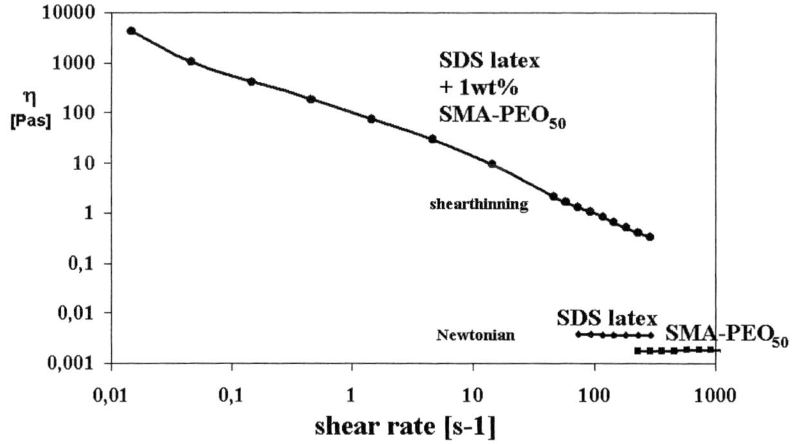


*Fig.9. Viscosity of the SDS latex, a SMA-PEO$_{50}$ solution and the SDS latex plus 1wt.-% of SMA-PEO$_{50}$*

In Fig.9, the effect of the strongly hydrophobized SMA-PEO$_{50}$ is demonstrated. This surfactant, despite its many stearyl groups, is still easily soluble in water. Water solutions are almost as fluid as water itself. But the latex plus little SMA-PEO$_{50}$ is extremely viscous at rest ( and at low shear rates), $10^6$ times the latex itself, and the factor is still $10^2$ at high shear rates, although the dispersion is strongly shear thinning.

That a polymer which does not form a hydrogel in water, nonetheless, aggregates in the latex to a physical chain network, thickening it extremely, but without flocculating it, is spectacular. A mechanism of *associative thickening* that is known from other thickeners seems to apply[3,4]. As shown in Fig.10, the PEO branches of the surfactant chains are partly adsorbed on the particle surfaces and are tied together in a physical network by micelles formed by the hydrophobic parts, i.e. the backbone and the stearyl substituents.

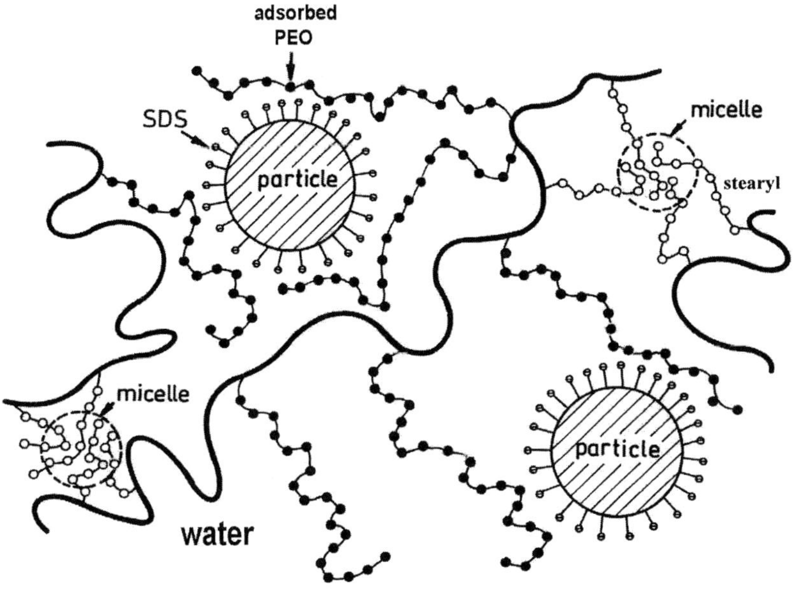


*Fig. 10. Associative thickening of a latex: chain network of graft copolymer surfactants as SMA-PEO$_x$, PEO branches partly adsorbed at the particle surface, physical network through hydrophobic micelles of the stearyl substituents*

# 3 Conclusions

The thermoplastic SMA, majorly consisting of styrene units but strongly functionalized with maleic anhydride units, proved to be a very useful source of polymeric surfactants. Anionic, cationic and nonionic surfactants could be prepared. These derivatives proved competitive as dispersants and thickeners. They cannot be used as flocculants, the molecular weight being too small.

SMA is a thermoplastic that can be processed in the melt by all methods of thermoplastic processing. Therefore, syntheses of the surfactants in the melt are presently being developed. While this is difficult in case of the ionic derivatives, it seems successful with the nonionic derivatives and the hydrophobisation.

***Acknowledgement***

We thank DSM, in particular Dr. M. van der Sanden, for the SMA copolymer and the support of this study.

***References***


1 C.L. McCormick, J. Bock, D.N. Schulz, Enc.Polym.Sci.Eng. **17** (1989) 730

2 T.C. Patton, *Paint Flow and Pigment Dispersion*, Wiley-Interscience, New York 1964

3 B.J. Cabane, J.Phys.Chem. **81** (1977) 1639

4 D.J. Lundberg, J.E. Glass, R.R. Eley, Polym.Mater.Sci.Eng. **61** (1989) 533; **67** (1992) 280

5 J. Bratby, *Coagulation and Flocculation*, Upland Press, Croydon UK, 1980

# Control of the Molecular Weight of Hyperbranched Polyglycerols

Holger Kautz, Alexander Sunder, Holger Frey*

Freiburger Materialforschungszentrum F M F und Institut für Makromolekulare Chemie der Albert-Ludwigs-Universität, Stefan-Meier Str. 21/31, D-79104 Freiburg, Germany

SUMMARY: Hyperbranched polyglycerols exceeding $M_n$ > 15,000 g/mol ($DP_n$ > 270) have been prepared, using a modified version of the synthetic protocol reported earlier. In the optimized process polydispersities recorded by SEC remained narrow, with $M_w/M_n$ < 1.7 over a large molecular weight range. The drastic increase of viscositity in the course of the reaction, was found to be responsible for the increased fraction of cycles (MALDI-TOF MS) at high molecular weights.

## Introduction

Recently, intense theoretical efforts[1)] have paved the way for the preparation of hyperbranched polymers with controlled molecular weights and well-defined degree of branching[2)]. It has been known for a long time that polycondensation of $AB_m$ monomers leads to hyperbranched polymers possessing large polydispersity[3)] and does not permit control of molecular weights. An alternative strategy is based on the slow addition of suitable latent $AB_m$ (mostly $AB_2$) monomers, which has recently been employed to control molecular weights and to lower the polydispersity of hyperbranched polymers. We reported on the controlled ring opening multibranching polymerization (ROMBP) of glycidol, representing a latent cyclic $AB_2$ monomer[4)]. The resulting polyether-polyols (Figure 1) possess polydispersities below 1.5. These flexible hyperbranched polyethers have proven useful polymeric precursors for the synthesis of amphiphilic molecular nanocapsules[5)] as well as for a new type of hyperbranched liquid crystalline polymers[6)]. However, to date, molecular weights of polyglycerol and respective copolymers have been limited to 6,000 g/mol ($DP_n$ < 90), since the addition of larger amounts of monomer to the initiator-core had not been explored. In this report, we demonstrate that also higher molecular weight materials are accessible after modification of the synthetic procedure. In addition, a reactor with outlet valve permitted to take samples in

 CCC 1022-1360/00/$ 17.50+.50/0

the course of the polymerization under slow-monomer-addition conditions, allowing us to monitor polymer formation and potential side reactions in detail.

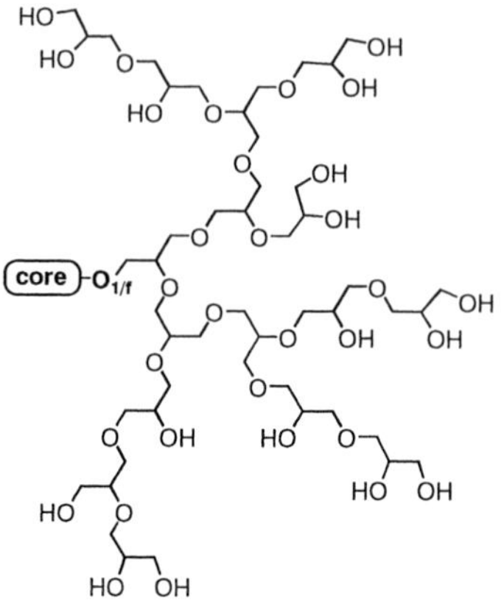


Fig. 1: Schematic structure of a hyperbranched polyglycerol consisting of dendritic, linear and terminal monomer units[9]

## Synthesis and Characterization

The basic synthetic protocol has been described previously[4] and is based on the slow-monomer-addition principle: glycidol is added slowly and continuously to the partially deprotonated polyol initiator, using a dosing pump. A fast protonation/deprotonation equilibrium ensures activity of all potential propagation sites of the growing polyfunctional macromolecules. Usually, monomer addition is complete after a few hours. The slow-monomer-addition strategy results in a very low monomer concentration actually present in the reaction mixture. Thus, exclusive reaction of the monomer with the growing polyfunctional macromolecules occurs, resulting in controlled molecular weights.

We were able to scale-up the reaction using a 2l reactor which enables us to produce hyperbranched polyglycerols on a scale of 500 g within one day. Furthermore, an outlet valve permits to take samples in the course of the reaction. This is crucial to monitor the polymerization process in detail to optimize reaction parameters, such as stirring intensity, stirrer geometry and dilution.

Applying the same protocol as reported for the lower molecular weight samples (< 6,000 g/mol)[4] resulted in considerably broader molecular weight distributions for higher monomer/initiator ratios ($M_w/M_n$ > 4; Figure 2, Table 1). In addition, SEC plots showed a bimodal molecular weight distribution with a shoulder in the lower molecular weight region (circled area in Figure 2).

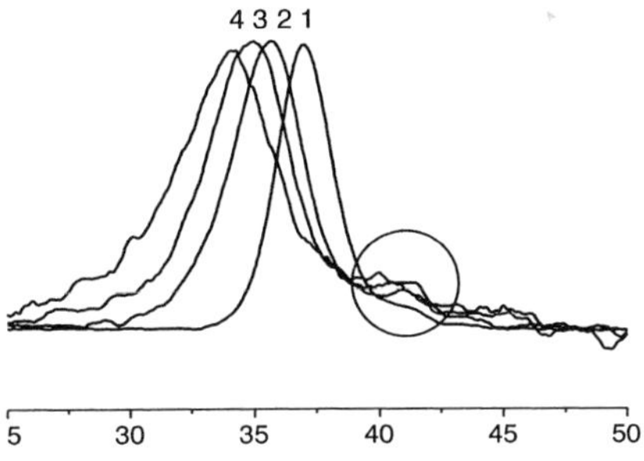


Fig. 2: SEC traces of hyperbranced polyglycerol samples taken from the same (non-optimized) reaction (measured in DMF, 45 °C).

Table 1. Characterization data of hyperbranched polyglycerols using bis(2,3-dihydroxypropyl)hexadecyl amine as initiator.

| # | theor. $M_n$ [a] | $M_w/M_n$ [b] | exp. $M_n$ [b] |
|---|---|---|---|
| 1 | 7,000 | 1.96 | 7,000 |
| 2 | 9,000 | 3.01 | 8,200 |
| 3 | 11,000 | 4.89 | 17,000 |
| 4 | 16,000 | 4.65 | - |

a) calculated from monomer/initiator ratio
b) measured by SEC in DMF using poly(propylene oxide) calibration

Both the bimodality of higher molecular weight samples as well as the general broadening of the main distribution mode are undesired effects, explained by limitations peculiar to the slow-addition procedure employed. Mixing of the reactants is a crucial issue for polymerizations carried out under slow-monomer-addition conditions. Under the reaction conditions reported previously, the viscosity increases drastically, reducing the accessibility of all propagation sites. A major prerequisite for a controlled multibranching polymerization is the reversible termination, i. e., due to intra- and intermolecular transfer all propagation site (i.e., hydroxyl groups) remain active. Thus, the broadening of the molecular weight distribution is explained by the increasingly impeded diffusion of the monomer to the active sites slowing down the reversible termination continuously. When this happens, self condensing steps that may eventually lead to intramolecular cyclization can become a side reaction competing with the initiated growth process. As a consequence, oligomers with polymerizable core unit are formed. Both reaction of these species with other polymer molecules as well as intramolecular cyclization are expected to lead to a broadening of the molecular weight distributions, similar to "classic" $AB_2$ polycondensations.

By modification of the process parameters, i.e. increasing stirring intensity and changing stirrer geometry we were able to achieve more efficient mixing. In addition to a reduction of the monomer addition rate we introduced diethylene glycol dimethyl ether (diglyme) as an inert and emulsifying solvent (polyglycerols of higher molecular weight are not soluble in diglyme) in order to dilute the monomer and reduce viscosity of the polymerization mixture We were thus able to obtain hyperbranched polyglycerols with monomodal, narrowed molecular weight distributions as well as molecular weights closer to the theoretical values. For example, as demonstrated in Figure 3 and Table 2, $M_w/M_n$ was lowered by a factor of 2 for $M_n$ = 9,000.

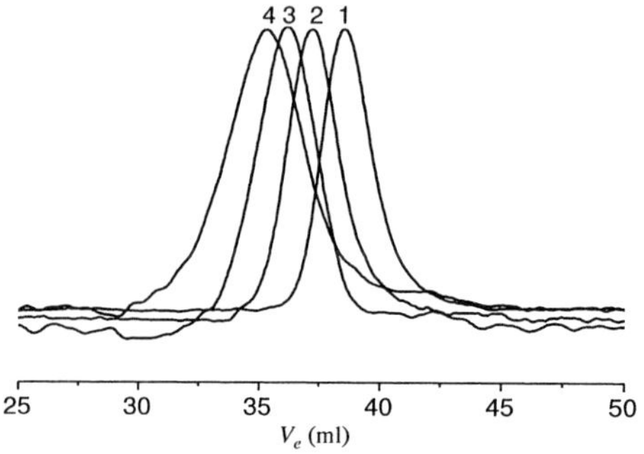


Fig. 3: SEC traces of various hyperbranched polyglycerol samples taken from the same reaction batch (optimized reaction parameters, measured in DMF at 45°C)

Table 2. Characterization data of hyperbranched polyglycerols using bis(2,3-dihydroxypropyl)hexadecyl amine as initiator.

| # | theor. $M_n$ [a)] | $M_w/M_n$ [b)] | exp. $M_n$ [b)] |
|---|---|---|---|
| 1 | 2,400 | 1.62 | 2,700 |
| 2 | 5,300 | 1.80 | 6,000 |
| 3 | 9,000 | 1.65 | 12,000 |
| 4 | 17,000 | 2.99 | 20,000 |

a) calculated from monomer/initiator ratio
b) measured by SEC in DMF using poly(propylene oxide) calibration

In order to gain insight into potential side reactions, MALDI-TOF spectra have been recorded. It has to be noted that only a low molecular weight fraction of the whole molecular weight distribution can be detected by MALDI-TOF MS. However, since cycles are known to be present mainly in this fraction, as shown for hyperbranched polyesters[8)], MALDI-TOF is a useful tool to determine whether cycles have been formed during the polymerization. Figure 4 a-c shows the MALDI-TOF spectra of the three samples taken from the same reaction after different portions of glycidol monomer had been added. Figure 4a gives the spectrum of a hyperbranched polyglycerol sample with $M_n$ = 7,000 g/mol. Clearly, there is no subdistribution and the signals evidence 100% initiated (core-bearing) polymer molecules. When viscosity increases with molecular weight the formation of cycles is observed in the low molecular weight fraction, which is confirmed by a subdistribution in the MALDI-TOF spectrum (Figure 4 b). Calculations confirm that the newly appearing peaks belong to cyclic species bearing one potassium as counter ion ($M = DP*74.1 + M(K^+)$ ) (Table 3).

Interestingly, the high viscosity in the final stages of the reaction even leads to a reversal of the intensities of cyclic and initiated species (Figure 4 c). Once more, it should be emphasized that the main distribution mode containing the desired high molecular weight material can not be observed in this MALDI-TOF spectrum. The fraction of cycles revealed by MALDI-TOF is small when compared to the whole batch and could not be observed in SEC.

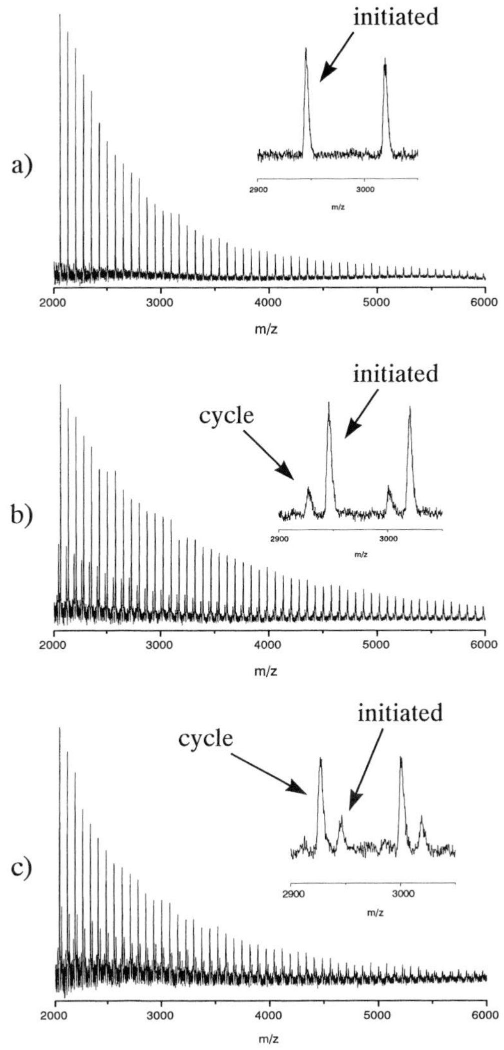


Fig. 4 a-c: MALDI-TOF MS spectra of three polyglycerol samples with increasing molecular weight taken from the same reaction batch, starting with a) $M_n$ = 7000 (molar ratio glycidol : TMP = 95:1). Inserts display enlarged spectrum from $m/z$ = 2,900 to 3,050 g/mol.

Table 3. MALDI-TOF $m/z$ of cycles and initiated species (initiator: bis(2,3-dihydroxypropyl)hexadecyl amine)

| | [cyclic species·$K^+$] $m/z$ | | [acyclic species·$K^+$] $m/z$ | |
|---|---|---|---|---|
| $DP_n$ | calcd. | found | calcd. | found |
| 34 | | | 2947.4 | 2945.5 |
| 35 | | | 3021.5 | 3020.5 |
| 39 | 2928.2 | 2928.0 | | |
| 40 | 3002.3 | 3001.7 | | |

Materials properties change distinctively when comparing lower and higher molecular weight samples of hyperbranched polyglycerols. Whereas samples with molecular weight of a few thousand g/mol were obtained as honey-like, sticky oils, polyglycerols with $M_n$ exceeding 10,000 g/mol are almost non-sticky, highly viscous materials that do not flow. Detailed rheological and mechanical characterization is in progress.

Absolute molecular weight determination of hyperbranched polyglycerols with $M_n > 10{,}000$ g/mol represents a challenge, since the commonly employed characterization methods reach their limits. NMR only gives reliable results for polyglycerols with molecular weight < 10,000 g/mol. The accuracy of the previously employed vapor pressure osmometry also decreases with molecular weight, especially for the highly polar polyglycerols. Nevertheless, with respect to experimental error, measurements with fully acetylated (hydrophobized) samples confirm the high molecular weights expected from the monomer/initiator ratio and SEC results.

## Conclusions

We have been able to enhance molecular weight of hyperbranched polyglycerols up to 20,000 g/mol ($DP_n$ = 270). Undesired cyclization leading to a broadening of the polydispersity and loss of control could be suppressed over a large molecular weight range by optimization of the polymerization process. The obtained samples, with up to 270 hydroxyl end groups, can be used as reactants for a wide range of possible applications. Currently, a detailed investigation of structure-property correlations of these high molecular weight hyperbranched polyglycerols and their intriguing rheological properties is in progress.

## References


1. a) W. Radke, G. Litvinenko, A. H. E. Müller, *Macromolecules* **31**, 239 (1998); b) R. Hanselmann, D. Hölter, H. Frey, *Macromolecules* **31**, 3790 (1998)
2. Overview of synthetic strategies for the preparation of hyperbranched polymers: A. Sunder, J. Heinemann, H. Frey, *Chem. Eur. J.* (2000), in press
3. a) P. J. Flory, *J. Am. Chem. Soc.* **74**, 2718 (1952); b) B. I. Voit, *Acta Polymer*. **46**, 87 (1995);
4. A. Sunder, R. Hanselmann, H. Frey, R. Mülhaupt, *Macromolecules* **32**, 4240 (1999)
5. A. Sunder, M. Krämer, R. Hanselmann, R. Mülhaupt, H. Frey, *Angew. Chem.* **111**, 3758 (1999); *Angew. Chem. Int. Ed.* **38**, 3552 (1990)
6. A. Sunder, M.-F. Quincy, R. Mülhaupt, H. Frey, *Angew. Chem.* **111**, 3107 (1999); *Angew. Chem. Int. Ed.* **38**, 2928 (1990)
7. a) P. Bharati, J. S. Moore, *J. Am. Chem. Soc.* **119**, 3391 (1997); b) P. Bharati, J. S. Moore, *Macromolecules* (2000), *in press*.
8. A. Burgath, A. Sunder, H. Frey, *Macromol. Chem. Phys.* (2000), in press
9. A. Sunder, R. Mülhaupt, R. Haag, H. Frey, *Macromolecules* **33**, 253 (2000)

# Hyperbranched Polymers with a Degree of Branching of 100%

Gerhard Maier*, Christina Zech, Brigitte Voit[a], Hartmut Komber[a]

Lehrstuhl für Makromolekulare Stoffe, Technische Universität München, Lichtenbergstrasse 4, D-85747 Garching, Germany
[a] Institute for Polymer Research Dresden, e.V., Hohe Strasse 6, D-01067 Dresden, Germany

SUMMARY: 4-Methoxy-4'-(3''-N-maleimidopropoxy)benzaldazine has been prepared and used for the synthesis of hyperbranched polymers by ‚criss-cross'-cycloaddition. In this type of reaction two unsaturated groups, such as the maleic imide groups, are added to the azine function. Hence, the azine behaves as a difunctional system, and therefore 4-Methoxy-4'-(3''-N-maleimidopropoxy)benzaldazine is an $AB_2$ monomer, with the maleic imide group being the A-function, and the azine group representing the $B_2$ part. The nature of this cycloaddition reaction ensures the coupled reaction of the two B-functions of each monomer unit. Consequently, the resulting polymer consists exclusively of branched and terminal repeating units and therefore has a degree of branching of 100%.

## Introduction

Hyperbranched polymers have attracted much interest recently because of their unique set of properties. Like dendrimers, they possess e.g. an extremely large number of functional groups, a higher segment density than linear analogs, good solubility and low melt viscosity. In contrast to dendrimers, hyperbranched polymers are not spherical, they are not unimolecular, their architecture is quite uncontrolled, and the functional groups occur throughout the molecule rather than exclusively at the surface. However, hyperbranched polymers are easily prepared from monomers of the $AB_x$-type in a straightforward one-step synthesis.

A number of approaches have been proposed to achieve control of the degree of branching of hyperbranched polymers in order to shift the properties more closely to that of dendrimers. Examples are slow monomer addition (with or without the addition of core molecules)[1,2,3,4], the addition of core molecules in self-condensing vinyl polymerization[5,6,7,8,9], the use of monomers with preformed branch points[10,11], activation of the second B-group by reaction of the first B-group[12,13,14], and post-synthetic modification of the linear units of conventional hyperbranched polymers with monofunctional compounds into branched units[15].

 CCC 1022-1360/00/$ 17.50+.50/0

While all these methods have been shown indeed to have an influence on the degree of branching, none of them excludes the formation of linear repeating units completely, and therefore none of these methods guarantees definitely the formation of hyperbranched polymers with 100% branching. In this paper we are presenting a different approach to this problem.

## Selection of a suitable type of reaction

Complete branching of an $AB_2$ monomer must occur, when both B groups are coupled in a way which prevents reaction of a single B group and enforces reaction of both B groups simultaneously. The following requirements are necessary for a reaction or reaction sequence in order to show this type of behavior:

- The reaction of the first B group with an A group must result in an *unstable* intermediate.
- The formation of the intermediate must be completely *reversible*.
- The reaction of the second B group with another A group must convert the intermediate into a *stable* final product.
- This second step must be *irreversible* under the reaction conditions.
- No other reaction or reaction sequence must occur.

The 'criss-cross'-cycloaddition reaction[16,17] between azines and unsaturated compounds meets these criteria. As Huisgen[18,19] proposed and Burger[20,21] proved, this reaction proceeds in two steps via an azomethinimine intermediate (Scheme 1). Usually, this intermediate is unstable, and unless the second step of the reaction occurs immediately after the first step to conclude the sequence under formation of the final stable product, the intermediate decomposes, reversing its formation reaction. Scheme 1 shows the mechanism using the reaction between a benzaldazine derivative and an isocyanate as an example.

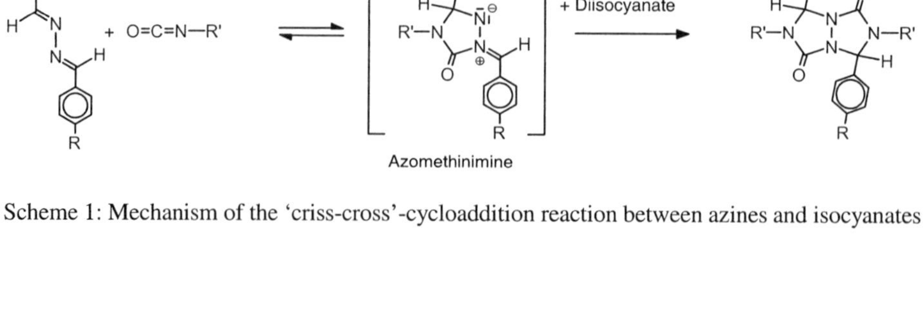


Scheme 1: Mechanism of the 'criss-cross'-cycloaddition reaction between azines and isocyanates

As we have reported previously[22,23,24,25,26,27)], the sequence shown in Scheme 1 reliably leads to telechelic polymers with two isocyanate end groups, when an azine is reacted with a diisocyanate. The functionality of these telechelics is independent of the stoichiometry of the starting materials. Thus, it is clear that the azomethinimine can follow only two reaction paths: either it adds another isocyanate group, or, if none is available, it decomposes into azine and isocyanate. Consequently, the azine group either reacts as a difunctional unit, or it does not react at all. Based on this observation, an azine which contains an unsaturated function such as an isocyanate or a CC double bond represents an $AB_2$ monomer according to the requirements discussed initially.

## Synthesis of the monomer

Potential monomer structures are shown in Scheme 2. They are asymmetrical azines, which are not easy to synthesize in high yield and sufficient purity for polycondensation reactions..

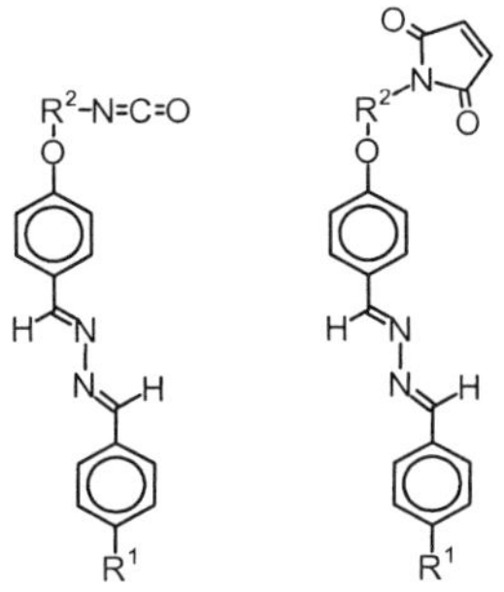


**1** **2**

Scheme 2: Target structures

Azines are usually prepared from carbonyl compounds and hydrazine under evolution of water. They undergo exchange reactions easily in the presence of water. Consequently, the preparation of asymmetric azines directly from different carbonyl compounds and hydrazine is often unsuccessful, since normally one of the symmetrical azines crystallizes more readily than the asymmetrical derivative and is therefore isolated first. As the most readily isolated derivative is removed from the equilibrium by crystallization, all of the asymmetrical azine is by and by converted to the symmetrical ones, if no precautions are taken. Many approaches can be found in the literature to overcome this problem. For our monomers, it proved to be most useful to prepare the hydrazone of one of the molecule halvesA and the Schiff-base with aniline of the other ‚halve' (Scheme 3). On reaction of these two, aniline is formed rather than water, and at exact stoichiometry the desired asymmetrical azine is obtained exclusively. Scheme 3 shows the reaction sequence for monomer **2a** ($R^1$ = -$OCH_3$, $R^2$ = -$(CH_2)_3$-).

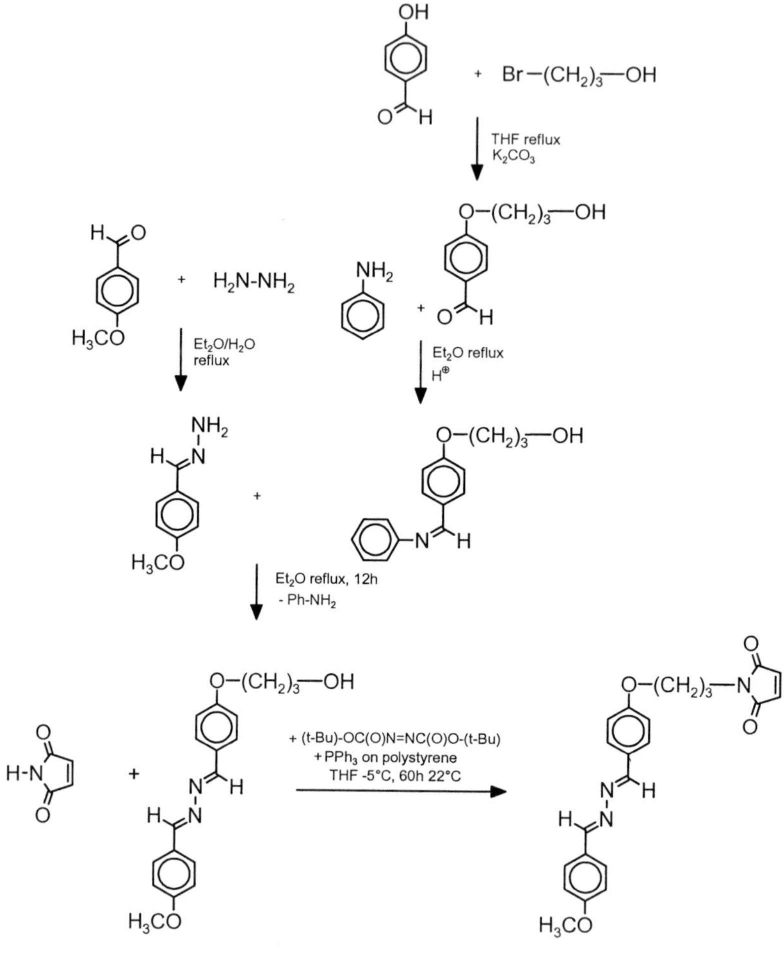


**2a**

Scheme 3: Synthesis of monomer **2a**

From the pathway shown in Scheme 3 it is clear that rather strong nucleophiles such as amine and hydrazone groups are involved. These would react with the isocyanate function of monomers of type **1**, which demands the use of a protective group for the isocyanate. So far, our attempts to prepare the asymmetrical azine with an isocyanate group in sufficiently pure form have not been successful.

In contrast, the maleic imide group is less readily attacked by nucleophiles, and thus seemed more promising. Finally, the pathway shown in Scheme 3 provided access to the desired monomer **2a** in pure form. A mass spectrum of the product as obtained from the last synthesis step indicated the absence of the two symmetrical azines. Even traces of these would result in undesired structures: One of the symmetrical derivatives acts as an $A_2$-monomer, introducing linear repeating units; the other one acts as an $A_2B_2$-monomer, causing crosslinking.

## Polymerization

Monomer **2a** was heated to 180-185 °C to achieve polymer formation. It is interesting to note that products which were formed under vacuum or in presence of air (moisture!) were indeed partly insoluble due to crosslinking. This behavior should be expected based on the characteristic reactivity of asymmetrical azines. In the presence of traces of water, both symmetrical azines are formed from the asymmetrical azine in monomer **2a**. Since one of them is actually an $A_2B_2$-monomer, crosslinking is to be expected under these conditions. Also, when the reaction is conducted under vacuum, both symmetrical azines are formed, with the sublimation of 4-methoxybenzaldazine being the most probable driving force. In contrast, soluble polymers were obtained under dry nitrogen atmosphere. Table 1 summarizes the results.

Table 1: Polymerization of monomer **2a**

| No. | Reaction temperature °C | Reaction time min | Atmosphere | $\overline{M}_n$ a) $g \times mol^{-1}$ | $\overline{M}_w$ a) $g \times mol^{-1}$ |
|---|---|---|---|---|---|
| **1** | 185 | 15 | dry $N_2$ | 5700 | 12500 |
| **2** | 185 | 30 | dry $N_2$ | 4400 | 8600 |
| **3** | 180 | 15 | air | insoluble | |
| **4** | 185 | 15 | vacuum | insoluble | |

a) GPC in $CHCl_3$, calibration with linear polystyrene

*Molar Masses*

The molar masses are not yet high enough to show the typical multimodal distribution of hyperbranched polymers. The reason for the limited molar masses obtained so far has to be clarified in the future. Possibly, the reaction conditions can be improved. However, there is also the possibility of an inherent limitation due to the requirement of quasi-simultaneous reaction of both parts of the azine: as the number of maleic imide groups decreases with increasing conversion, the reaction may come to a stop because of the strongly decreasing probability of an encounter between an azine group and two maleic imide groups at the same time. It is quite obvious that the reaction is third order in functional group concentration, while conventional condensation reactions used for the preparation of hyperbranched polymers are second order in functional group concentration (provided external catalysis is used). Further work is required to shed light on this. Scheme 4 shows the GPC curves as obtained with RI-detection and polystyrene calibration (a), and a light scattering detector (b).

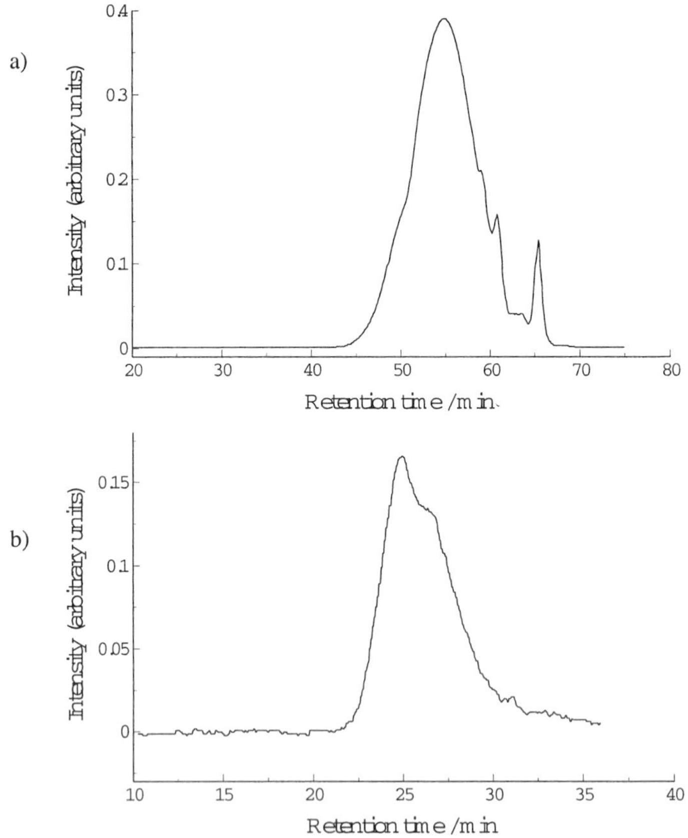


Scheme 4: GPC curves of the polymer from **2a** (entry 1, Table 1); a) in $CHCl_3$ with linear polystyrene calibration and RI detection; b) in THF with light scattering detection

*DSC*

DSC measurements of the soluble hyperbranched polymers exhibit a glass transition temperature of $T_g$ = 136 °C. No other phase transition were observed.

*Chemical structure and potential defects*

Scheme 5 shows the idealized structure of the hyperbranched polymers.

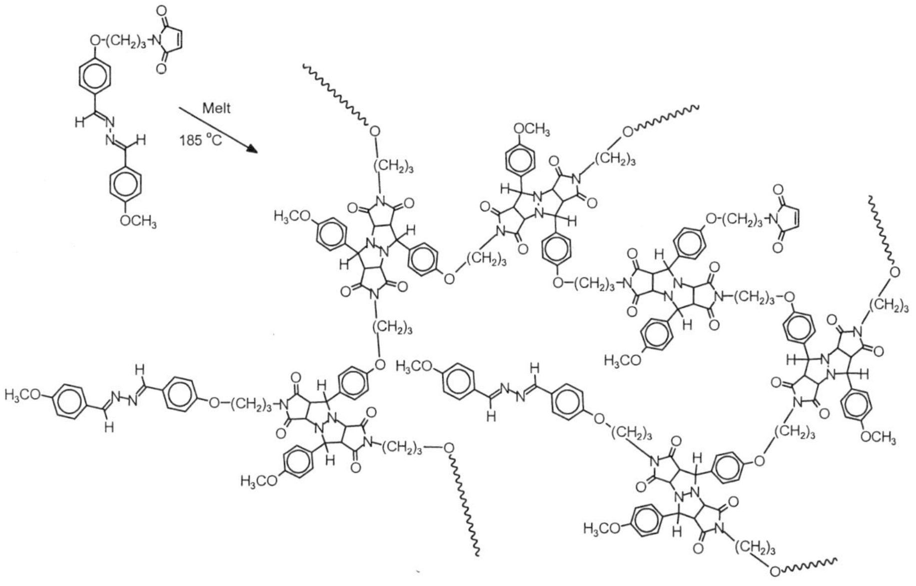


Scheme 5: Structure of the hyperbranched polymers

Any defect structures which might be caused by undesired side reactions would result in incomplete branching, and hence the original idea of a perfectly branched hyperbranched polymer would be compromised. Therefore, the analysis of the chemical structure and the search for defects is important.

The most obvious method to check for the presence of defect structures is MALDI-TOF mass spectroscopy. Perfect branching requires the addition of exactly two repeating units in every addition step, and therefore the difference between the masses of the oligomers must equal twice the mass of the monomer. Defects would result in the addition of only one repeating unit, resulting in a linear, non-branched unit, and should be easily discernible because of the resulting deviation of the molar mass. However, so far we have not been able to obtain a MALDI-TOF mass spectrum. The problem appears to be the absorption of the terminal azine units, which is close to the wavelength of the laser used for desorption, causing fragmentation.

Thus, NMR spectroscopy was used to check for defect structures. Scheme 6 shows the most probable defects.

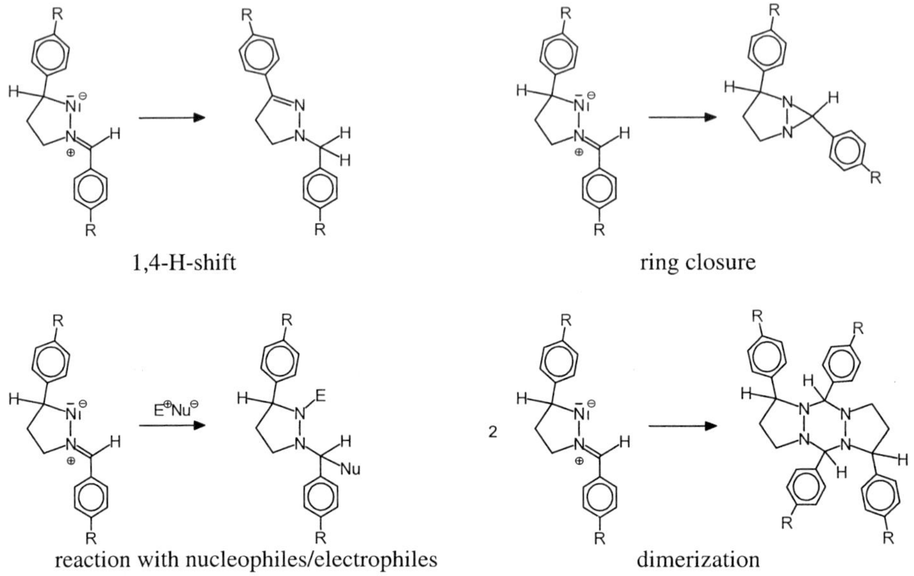


Scheme 6: Possible defect structures

The shift of a hydrogen atom from 1-position of the 1,3-dipole to the 4-position is comparable to the 1,5-H-shift of dienes, and it is frequently observed[18,19] for various 1,3-dipoles. However, it is a concerted orbital symmetry controlled reaction, and therefore it requires a cyclic transition state. In our case, this is impossible, because the azomethinimine is fixed in a `transoidA state, because it is part of the five-membered ring, and partly exocyclic.

1,3-Dipoles can form a three membered ring in an electrocyclic reaction. This can not be excluded here, but the high reaction temperature of 180-185 °C should prevent the formation of the strained three membered ring.

The cyclodimerization is unlikely due to the considerable steric crowding of the resulting structure.

Finally, the addition of nucleophiles could cause problems. Again, the high reaction temperature should help in part, because the presence of water in the reaction at 180 °C is not very likely. In addition, model reactions were performed, in which an azine was reacted with a maleic imide in the presence of phenol as a model nucleophile. The only product found in the resulting product was the cycloaddition product. No reaction with phenol was observed, even at molar concentrations of phenol. Thus, at least at high reaction temperatures the system appears to be quite robust, and undesired side reactions can obviously be avoided.

The carbon 13 NMR spectrum of the hyperbranched polymer **1** is shown in Scheme 7. It looks quite complicated at first sight, but all signals have been assigned by comparison with model compounds. Four isomers of the central tetracyclic ring system (see Scheme 5) have been identified[28]. Scheme 8 compares the spectra of a model compound (all isomers present), a linear model polymer, and the hyperbranched polymer. The similarity of the signal pattern clearly indicates the agreement between the chemical structures of the models and the hyperbranched polymer.

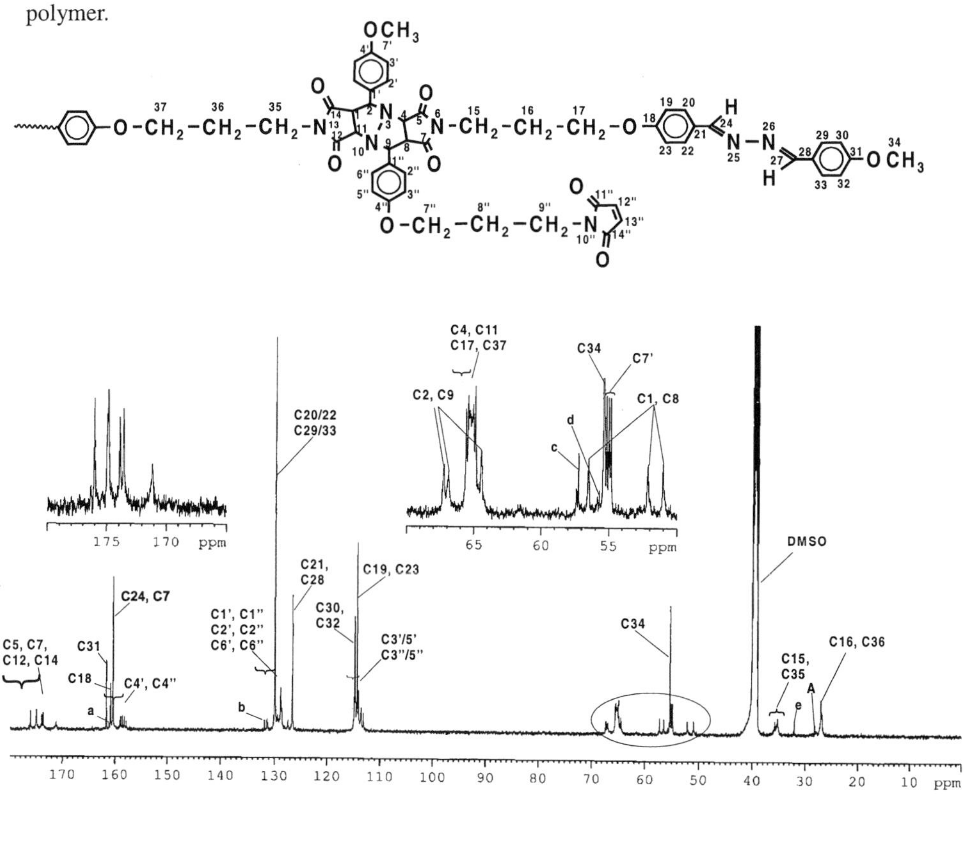


Scheme 7: $^{13}$C NMR of the hyperbranched polymer

2. linear model polymer

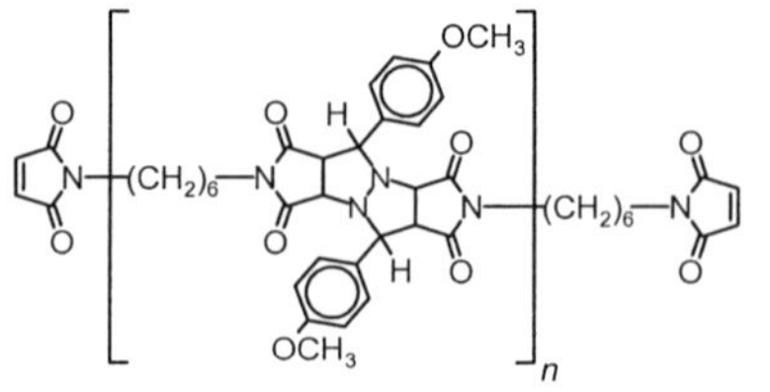


3. low molar mass model compound

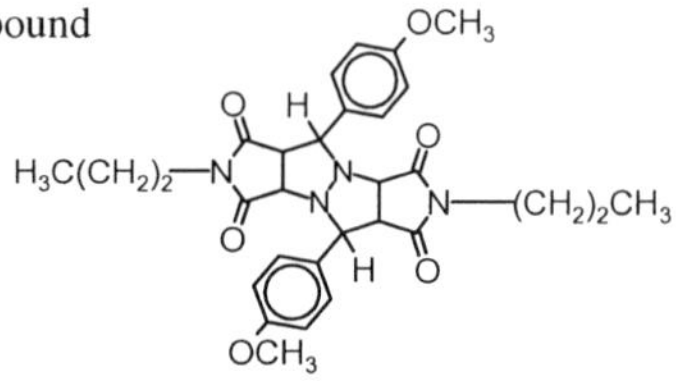


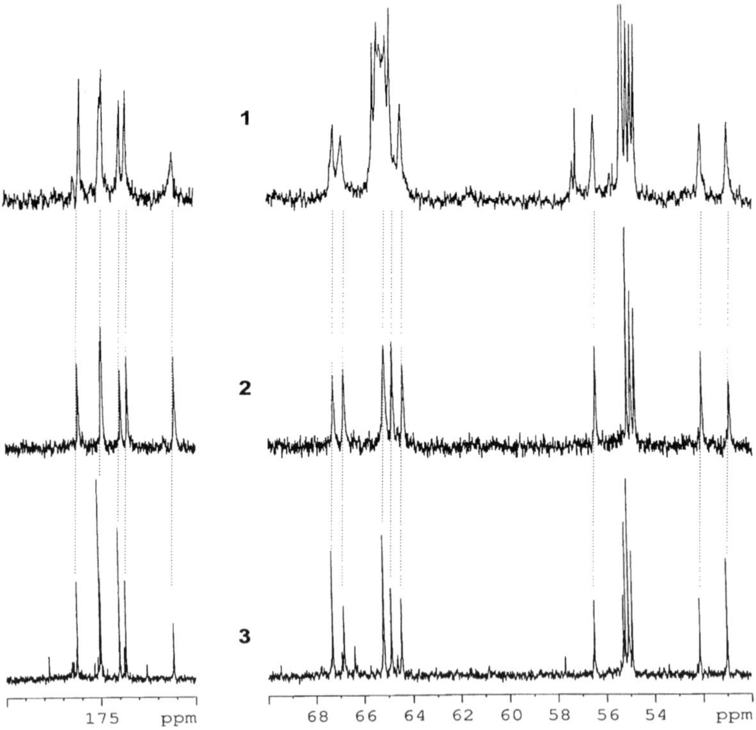


Scheme 8: Comparison of the $^{13}C$ NMR spectra of 1) the hyperbranched polymer, 2) a linear model polymer, 3) a low molar mass model compound

**Conclusions**

An $AB_2$ monomer including an azine group to serve as the two B-functions and a maleic imide group as the A-function was successfully synthesized. Its polymerization at 180-185 °C resulted in the formation of a perfectly hyperbranched polymer. No indication of any defect structures, leading to linear repeating units, were found by NMR spectroscopy or model reactions. Further characterization and improvements of the molar mass of the polymers to determine whether they are closer to perfect dendrimers in their physical properties than conventional hyperbranched polymers is going on.

## References


1. D. Hölter, H. Frey, *Poly. Mat. Sci. Eng.* **77**, 165 (1997)
2. A. Burgath, R. Hanselmann, D. Hölter, H. Frey, *Polym. Mat. Sci. Eng.* **77**, 166 (1997)
3. D. Hölter, H. Frey, *Acta Polymer.* **48**, 298 (1997)
4. W. Radke, G. Litvinenko, A. H. E. Müller, *Macromolecules* **31**, 239 (1998)
5. E. Malmström, F. Liu, R. H. Boyd, A. Hult, U. W. Gedde, *Polym. Bull. (Berlin)* **32**, 679 (1994)
6. E. Malmström, M. Johansson, A. Hult, *Macromolecules* **28**, 1698 (1995) (The extremely high degrees of branching reported herein are due to inappropriate assignment of NMR signals, but values of the order of DB. 60% are realistic)
7. H. R. Kricheldorf, G. Löhden, *Macromol. Chem. Phys.* **196**, 1839 (1995)
8. H. R. Kricheldorf, O. Stöber, D. Lübbers, *Macromolecules* **28**, 2118 (1995)
9. W. J. Feast, N. M. Stainton, *J. Mater. Chem.* **5**, 405 (1995)
10. C. J. Hawker, F. Chu, *Macromolecules* **29**, 4370 (1996)
11. E. Malmström, M. Trollsås, C. J. Hawker, *Polym. Mat. Sci. Eng.* **77**, 151 (1997)
12. L. J. Hobson, W. J. Feast, A. M. Kenwright, *Polym. Mat. Sci. Eng.* **77**, 220 (1997)
13. L. J. Hobson, W. J. Feast, *Chem. Commun..* **1997**, 1877
14. L. J. Hobson, W. J. Feast, *Chem. Commun..* **1997**, 2067
15. C. Lach, H. Frey, *Macromolecules* **31**, 2381 (1998)
16. J. R. Bailey, N. H. Moore, *J. Am. Chem. Soc.* **39**, 279 (1917)
17. J. R. Bailey, A. T. McPherson, *J. Am. Chem. Soc.* **39**, 1322 (1917)
18. R. Huisgen, *Angew. Chem.* **75**, 604 (1993)

19. R. Huisgen, *Angew. Chem. Int. Ed. Engl.* **2**, 565 (1963)

20. K. Burger et al., *Angew. Chem.* **86**, 481, 482, 483, 484 (1974)

21. K. Burger et al., *Angew. Chem. Int. Ed. Engl.* **13**, 474, 475, 476, 477 (1974)

22. O. Nuyken, G. Maier, K. Burger, *Makromol. Chem.* **189**, 2245 (1988)

23. O. Nuyken, G. Maier, K. Burger, *Makromol. Chem.* **190**, 623 (1989)

24. O. Nuyken, G. Maier, K. Burger, A. Serra i Albet, *Makromol. Chem.* **190**, 1953 (1989)

25. O. Nuyken, G. Maier, K. Burger, *Makromol. Chem.* **191**, 2455 (1990)

26. G. Maier, K. Losert, O. Nuyken, *Makromol. Chem.* **194**, 2413 (1993)

27. G. Maier, A. Fenchl, G. Sigl, *Macromol. Chem. Phys.* **198**, 137 (1996)

28. C. Zech, Dissertation, Technische Universität München, 2000

# Functionalization of Silica Particles towards Cationic Polyelectrolytes using Vinylformamide and 1,3-Divinylimidazolidin-2-one as Monomers

Torsten Meyer, Petra Rehak, Christian Jäger,[#] Ina Voigt, Frank Simon[§], Stefan Spange,*

Polymer Chemistry, Institute for Chemistry, Chemnitz University of Technology, Strasse der Nationen 62, D-09111 Chemnitz, Germany, e-mail: stefan.spange@chemie.tu-chemnitz.de

[#] Institute of Physics, Friedrich Schiller University of Jena, Max Wien Platz, Jena, Germany

[§] Institute of Polymer Research, Hohe Straße 6, D-01069 Dresden, Germany, http://www.ipfdd.de

**SUMMARY:** The use of vinylformamide and 1,3-divinylimidazolidin-2-one, bisvinylethylene-urea (BVU), as well as of the poly(vinylformamide–vinylamine) copolymer (PVFA-co-PVAm) for silica surface functionalization has been investigated. Various procedures such as grafting from, crosslinking surface polymerization, and post-functionalization of adsorbed PVAm chains have been experimentally applied. The advantage of the different synthetic methods is discussed with regard to the resulting surface structure of the organic/inorganic hybrid materials.

## Introduction

Surface functionalization of inorganic substrates with polyelectrolytes plays an important role in technical application and material science[1-6]. Polyelectrolytes can be linked to polar surfaces of inorganic oxidic particles or flat substrates by different techniques: Either by a grafting from mechanism using covalently immobilized initiators or by consecutive adsorption of oppositely charged polyelectrolytes from water[1-7]. The two established surface structures are illustrated in Chart 1.

 CCC 1022-1360/00/$ 17.50+.50/0

Chart 1: Methods known for surface functionalization of inorganic substrates with polyelectrolytes.

multi layer adsorption of oppositely charged polyelectrolytes

grafted polyelectrolyte brush

Unfortunately, the use of water as solvent for surface functionalization limits the application of the polymers as well as chemical reaction for further derivatization. In this paper we will briefly show some adequate concepts which allow the construction of cationically active polyelectrolytes on silica particles surface. For this purpose, polyvinylamine (PVAm) as polymer has been chosen, because the primary amino groups offer a lot of possibilities for post-functionalizations on the hybrid particles surface. PVAm is synthesized by radically or cationically induced polymerization of vinylformamide and consecutive acid hydrolysis of polyVFA[8-10]. PolyVFA-co-PVAm is a promising polymer, due to the potentiality of controlling the polyelectrolyte properties widely, e.g. by the degree of hydrolysis (content of amino groups) and the pH strength of the aqueous solution (see Chart 2)[6]. The irreversibly linking to silica surfaces can be achieved by three different procedures:

- Grafting from of VFA with an immobilized initiator and hydrolysis of the grafted PVFA chains,
- surface crosslinking polymerization of VFA with BVU,
  and hydrolysis of the PVFA sections among the immobilized network, and
- post crosslinking of adsorbed PVFA-co-PVAm with bifunctional reagents (epoxides, isocyanates, carbon acid anhydride).

BVU can also be homopolymerized cationically on silica yielding a two-dimensionally ordered crosslinked polyurea network[11]. The use of the poly-BVU/silica hybrid particles for derivatization towards guanidine function is also described in this paper.

Chart 2: Compounds used for silica functionalization.

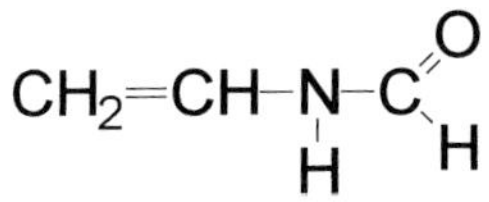

VFA

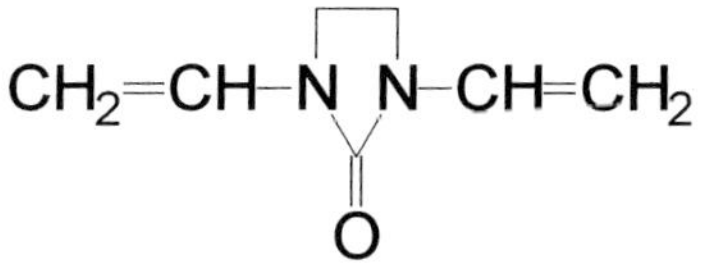

BVU

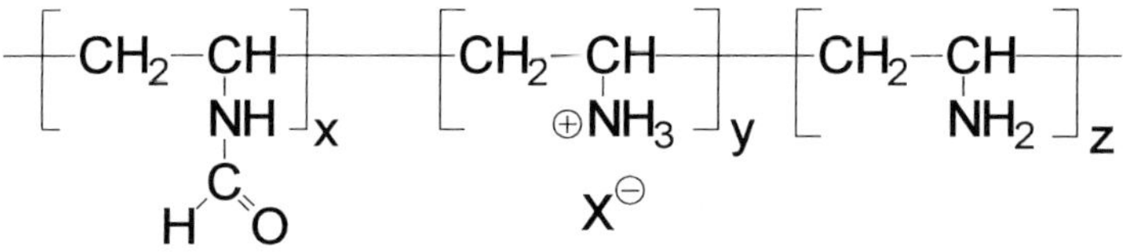

PVFA-co-PVAm

## Results and discussion

### Grafting from procedure

For the grafting from procedure, we used an azo compound as surface initiator as described by Rühe et al.[7]. This pathway for surface functionalization requires essentially three steps[12,13]:

1. The covalent immobilization of the surface initiator (often additional several steps),
2. the heterogeneously induced grafting from polymerization of the monomer in a slurry, and
3. the hydrolysis of the grafted PVFA chain towards PVAm (see Chart 3).

Of course, after each synthetic step, a purification by extraction is also necessary[7,11-13]. The structure of the surface bonded initiator is evidenced by solid state $^{13}$C-CP-TOSS-NMR spectroscopy (Figure 1) and quantitative elemental analysis.

Results for the surface grafting from reaction of VFA with the immobilized azo initiator on silica are given in Table 1.

Chart 3: Crafting from reaction of VFA.

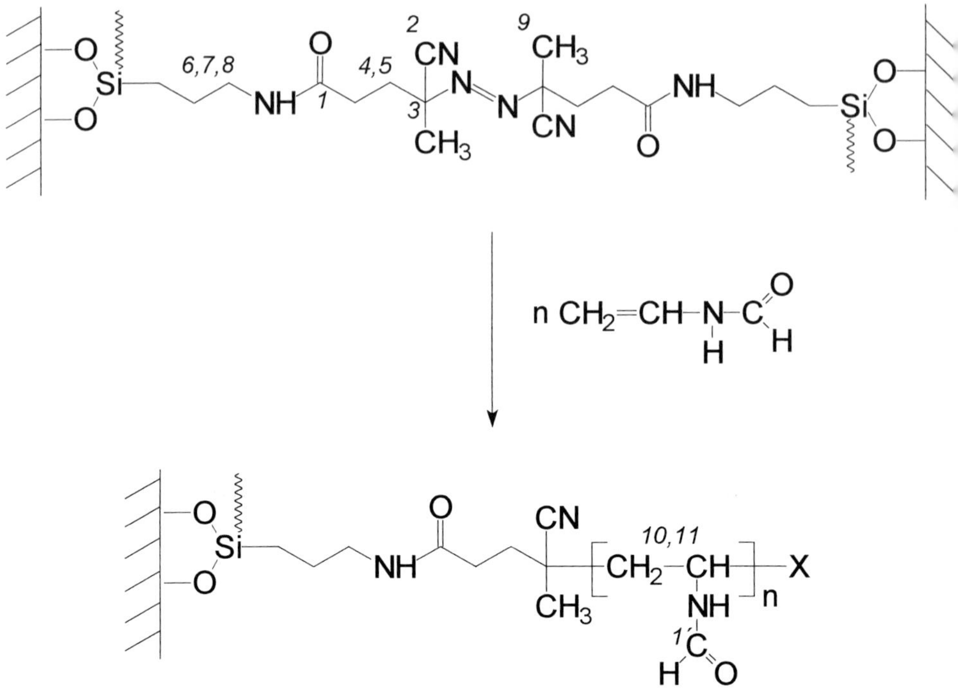


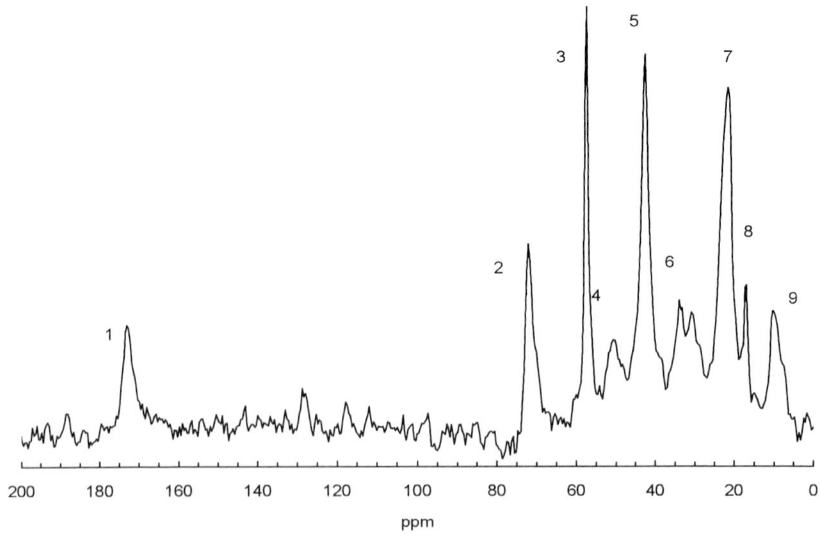


Fig. 1: Solid state $^{13}C\{^{1}H\}$CP-TOSS-NMR spectrum of the azoinitiator immobilized on a silica gel.

Table 1. Grafting yield and monomer conversion for the grafting from reaction of VFA with immobilized azo initiator on silica. T = 338.15 K.

| solvent | methanol | ethanol | benzene | toluene | water |
|---|---|---|---|---|---|
| reaction time (h) | 24 | 24 | 24 | 24 | 24 |
| conversion VFA (%) | 34 | 34 | 70 | 95 | 100 |
| graft density (g polym./g silica) | 0.2044 | 0.2057 | 0.2458 | 0.2770 | 0.1449 |

Unfortunately, PVAm is only soluble in water and, therefore, the PVFA formed during the grafting reactions in organic solvents adsorbs very strongly on the particle surface and limits the grafting reaction molecular weight and monomer conversion. In water, the monomer conversion is complete, but then the solution polymerization is preferred, because one part of the azo initiator initiates the solution polymerization which proceeds very well in water. In conclusion, the grafting from method is suitable for PVFA functionalization of silica but not very effective. Figure 2 shows the solid state $^{13}$C-CP-TOSS-NMR spectrum of a PVFA/silica hybrid particle sample.

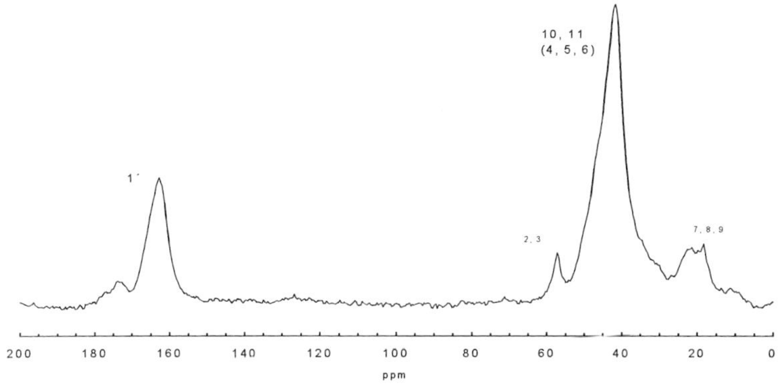


Fig. 2: Solid state $^{13}$C{$^{1}$H}CP-TOSS-NMR spectrum of grafted (from) PVFA on silica. Assignment of the signals see Chart 3.

A second problem results from the instability of the amide bond which links the grafted PVFA-chains to the silica surface, because these bonds can also be cleaved when the formamide groups are acidically hydrolysed to amino groups.

## Crosslinking surface polymerization

### *Cationic homopolymerization of BVU*

BVU can be cationically homopolymerized on silica and a crosslinked poly-BVU layer is formed[11].

However, polyurea is not yet a polyelectrolyte, because its basicity is reduced compared to PVAm[14]. But poly-BVU can be used for selective adsorption of metal ions ($Co^{2+}$, $Cu^{2+}$, $Ag^{+}$, $Au^{3+}$, $Fe^{3+}$) which offers a lot of further applications in chromatography and catalysis[14, 15].

An interesting option is the transformation of the urea unit into a guanidine structure, because guanidine is even a stronger base than a primary amine. This synthetic step requires the activation of the urea carbonyl bond with triphosgene[16]. The transformation of the urea moiety into a guanidine was studied with tert. butylamine as the reagent. This is shown in Figure 3 (inset: reaction scheme).

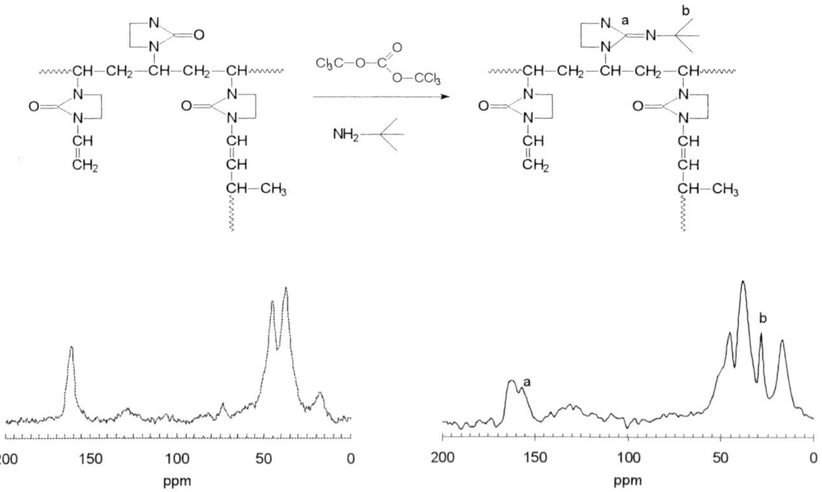


Fig. 3: Solid state $^{13}$C-CP-TOSS spectra for guanidine functionalization of poly-BVU network on silica. Inset: reaction scheme.

The introduction of the guanidine is clearly evidenced by the two new signals at $\delta = 157$ ppm and $\delta = 29$ ppm in the solid state $^{13}C\{^1H\}$-CP-TOSS-NMR spectrum, relating to the guanidine and methyl groups of the tert. butyl group, respectively. However, this modification reaction proceeds not to a complete conversion of the former urea units despite an excess of the functionalization reagent triphosgene/tert.butylamine is used. Other functionalization reactions with Lawessons reagent or reduction with complex hydride compounds have been found not successful for post-derivatization of poly-BVU. .

*Radical Copolymerization of VFA with BVU*

The target for this functionalization reaction is shown in Chart 4.

Chart 4: Radical surface crosslinking (core shell ) polymerization of VFA with BVU on silica particles and hydrolysis to PVAm.

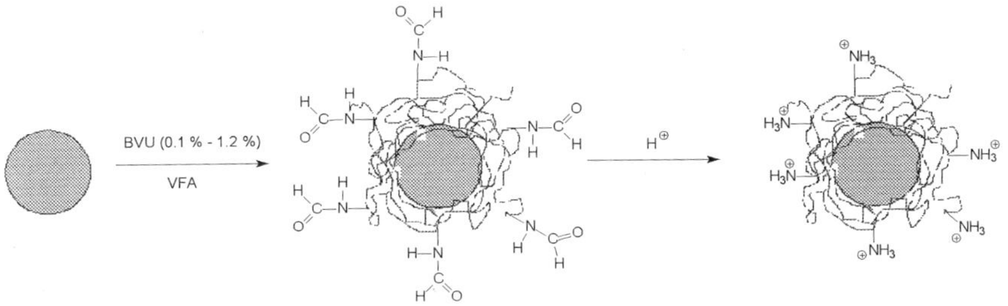


The content of the BVU component significantly determines the morphology of the PVFA-BVU gel formed among the silica matrices. If a concentration of about 1.5 % BVU is used, crosslinking between the silica particles is observed. Details on this reaction and the hydrolysis of the PVFA sections to PVAm will be reported in a following paper[17]. It should be briefly noted that the BVU units remain stable during the acid hydrolysation reaction. Therefore, this method is especially useful for immobilization high amounts of PVAm on silica.

**Adsorption of PVAm and post-crosslinking reaction**

PVAm can be adsorbed on silica from an aqueous solution[6]. However, the process is significantly dependent on the property of the PVFA-co-PVAm component. Best results (a high amount of PVAm is adsorbed) have been found when a PVAm sample with a high molecular weight ($M_n$ = 400.000 g $mol^{-1}$) and nearly complete hydrolysis of the former formamide groups (95 %) is used[6]. Surface charges of bare silica and PVAm/silica hybrids have been investigated by zetapotential and surface charge titration[6]. Based on these results, it was concluded that the adsorption energy of PVAm on silica in water is determined by electrostatic forces rather than by chemical interactions. Consequently, the PVAm layer can be easely removed by competing acids or bases. Irreversibly fixing of the adsorbed PVAm chains can be carried out with different bifunctional reagents, e.g. epoxides, diisocyanates[6], and fulleren molecules[18]. But this is not possible in water which is an essential solvent for PVAm[18]. Therefore, the synthetic concept requires two steps: adsorption of PVAm on silica in water, purification of the pre-functionalized PVAm/silica particles, and then post-functionalization in a suitable organic solvent. The synthetic steps are shown in Chart 5. Figure 4 shows the pH dependent zetapotential plots for bare silica, PVAm/silica, diisiocyanate functionalized PVAm/silica, and $C_{60}$ - functionalized PVAm silica.

Chart 5: Adsorption of PVAm on silica particles and post crosslinking reaction.

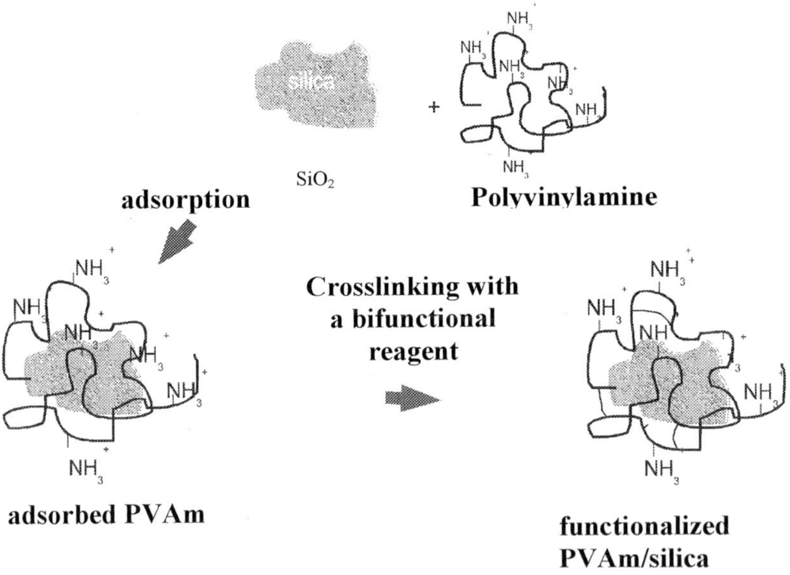


The results in Fig. 4 clearly show that post-functionalization does not disturb the polyelectrolyte properties of the crosslinked PVAm chains. However, a shift of the isoelectric point after post-functionalization to lower pH indicates a conversion of amino groups of the adsorbed PVAm chain.

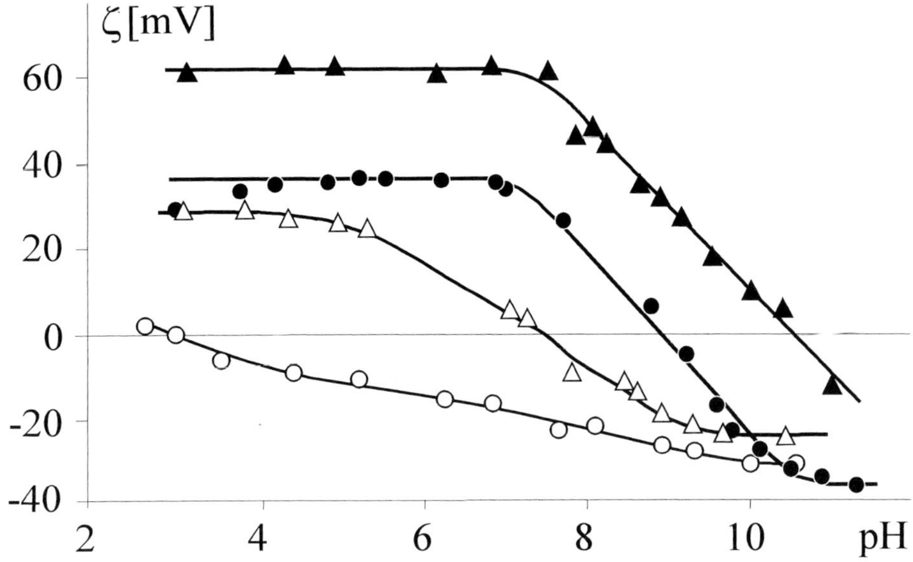


Fig. 4: Zetapotential as function of the pH for ○ bare silica, ▲ PVFA-*co*-PVAm-95/silica ($M_n$ = 400,000 g·mol$^{-1}$), ● PVFA-*co*-PVAm-95/silica functionalized with 0.10 wt: % (4,4′-diisocyanate)diphenyl methane, and Δ PVFA-*co*-PVAm-95/silica functionalized with $C_{60}$. All samples were measured in $1\cdot10^{-3}$ mol·l$^{-1}$ KCl at 25 °C[6].

But this effect is not so dramatic compared to bare silica. When PVAm is adsorbed on silica, a shift of the IEP of ΔIEP = 10 is observed. For details see ref.[6] and ref.[18].

## Conclusion

Vinylformamide is a very suitable monomer for surface functionalization, because the corresponding polymer which results after surface polymerization can be easely transformed by an acid hydrolysis to PVAm. Grafting from polymerization is possible but not very effective. Crosslinking copolymerization of VFA with BVU on silica is a suitable method. Adsorption of PVAm from water is also possible. This procedure requires a post-derivatization reaction in organic solvent. Experimental details are given in the references[6,11,14, 17, 18].

## Acknowledgement

Financial support by the DFG Schwerpunktprogramm "Polyelectrolytes" and scientific co-operation with the BASF, Aktiengesellschaft, are gratefully acknowledged.

## References

1. M. J. Roberts, G. A. Lindsay, *J. Am. Chem. Soc.* **120**, 11202 (1998)
2. C. Huguenard, J. Widmaier, A. Elaissari, E. Pefferkorn, *Macromolecules* **30**, 1434 (1997)
3. G. Decher, *Science* **227**, 1232, and references therein (1997)
4. G. B. Sukhorukov, E. Donath, H. Lichtenfels, E. Knippel, M. Knippel, H. Möhwald, *Colloids Surfaces* **A 137**, 253 (1998)
5. E. Donath, G. B. Sukhorukov, F. Caruso, S. A. Davis, H. Möhwald, *Angewandte Chemie* **110**, 2324, and references therein (1998)
6. I. Voigt, F. Simon, H. Komber, H. J. Jacobasch, S. Spange, *Colloid and Polymer Science* **278**, 48 (2000)
7. O. Prucker, J. Rühe, *Macromolecules* **31**, 592 (1998)
8. R. K. Pinschmidt, W. L. Renz, W. E. Caroll, K. Yacouth, J. Drescher, A. F. Nordquist, N. Chen, *J. Macromol. Sci. Pure Appl. Chem.* **A 34**, 1885 (1997)
9. R. J. Badesso, A. F. Nordquist, R. U. Pinschnmidt, D. J. Sagl, *Adv. Chem. Ser.* **248**, 489 (1996)
10. S. Spange, U. Eismann, A. Madl, J. Utecht, *Macromol. Chem. Rap. Comm.* **18**, 1075 (1997)
11. S. Spange, T. Meyer, *Macromol. Chem. Phys.* **200**, 1655 (1999)
12. R. Laible and K. Hamann, *Adv. Coll. Interf. Sci.* **13**, 65 (1980)
13. N. Tsubokawa, *Progr. Polym. Sci.* **17**, 417 (1992)
14. T. Meyer, S. Prause, S. Spange, *J. Coll. Interf. Sci.* in prep.
15. R. M. C. Santos, C. Airoldi, *J. Colloid Interf. Sci.* **183**, 416 (1996)
16. Organic Syntheses, Vol. **74** ed. by I. Shinkai, Notre Dame (1996) p. 101.
17. T. Meyer, S. Hesse, C. Jäger, C. Bellmann, S. Spange, *Coll. Polym. Sci.* in prep.
18. I. Voigt, F. Simon, M. Friedrich, S. Spange, *Langmuir*, submitted for

# Tailor-made membranes for alcohol/water pervaporation and ion separation prepared upon layer-by-layer adsorption of polyelectrolytes

Bernd Tieke*, Lutz Krasemann, Ali Toutianoush

Institut für Physikalische Chemie der Universität zu Köln,

Luxemburgerstr. 116, 50939 Köln, Germany

ABSTRACT: The paper is concerned with separating membranes prepared upon layer-by-layer adsorption of cationic and anionic polyelectrolytes on a porous substructure. The use of the resulting composite membranes for alcohol/water separation under pervaporation conditions, and for the separation of mono- and divalent ions is described. It is demonstrated that a suitable choice of the polyelectrolytes and the optimisation of the preparation conditions allow to tailor membranes with excellent separation capability.

## Introduction

Useful membranes are characterised by a high flux and a high selectivity. Either properties can be combined by preparation of so-called composite membranes consisting of a porous supporting membrane coated with a thin, homogeneous, pore-free separating layer[1,2]. The separating layer is usually prepared by casting from solution. Likewise, the separating layer can be prepared by alternating layer-by-layer adsorption of cationic and anionic polyelectrolytes[3] on porous or pore-free supporting membranes[4-6]. The method is schematically represented in Fig. 1. A porous substructure previously hydrophilised in $O_2$-plasma is dipped into a solution of a cationic polyelectrolyte, so that a thin layer of this compound is adsorbed and the surface charge is reverted. Subsequent dipping of the substrate into a solution of an anionic polyelectrolyte again leads to adsorption of a thin layer, and the surface charge is rendered negative again. Multiple repetition of the adsorption steps leads to a polyelectrolyte multilayer film, which has several advantages over separating layers cast from solution: Each adsorption step adds about 0.5 nm to the total thickness of the separating membrane[7], which allows to prepare ultrathin, homogeneous coatings of precisely controlled thickness. Moreover, a wide variety of polyionic compounds[3] can be adsorbed, which allow to tailor appropriate membranes for various separation problems. First experiments on the separation of gases[4-6], ions[8,9]

 CCC 1022-1360/00/$ 17.50+.50/0

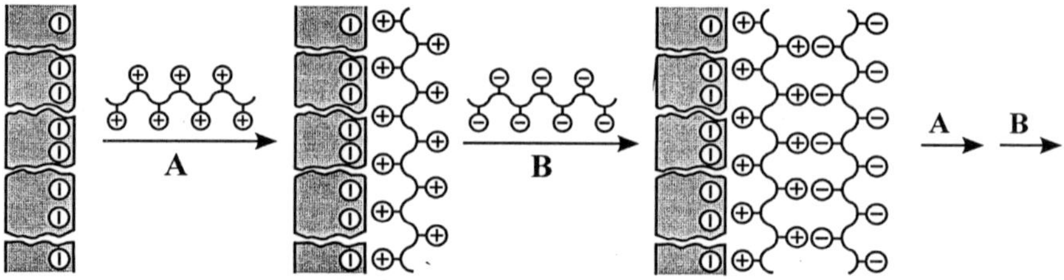


Fig. 1: Scheme of layer-by-layer adsorption of polyelectrolytes on plasma-treated porous supporting membrane. Multiple repetition of steps A and B leads to the ultrathin separating layer.

and liquid mixtures[6,9-13] have already been reported. Purpose of the present contribution is to demonstrate that upon a careful choice of the polyelectrolytes and an optimisation of the processing conditions polyelectrolyte multilayer membranes can be tailored, which are especially useful for distinct separation problems. This is demonstrated for the alcohol/water separation under pervaporation[14] conditions, and for the separation of mono- and divalent ions in aqueous solution. The separation behaviour and the influence of the polyelectrolyte structure and the processing parameters on flux and separation will be described.

## Pervaporation

Pervaporation[14] of the alcohol/water mixtures was carried out using the apparatus schematically represented in Fig. 2. The alcohol/water mixture ('feed solution') of constant temperature (T = 58,5 °C) was circulated from a 2 l reservoir to the pervaporation cell at a flow rate of 250 ml/min. The cell consists of a feed half cell and a permeate half cell with the membrane mounted in between. The permeate half cell was evacuated to a pressure $\leq 0.2$ mbar. The permeate was collected in the first condenser cooled with liquid nitrogen. After pervaporation for 2 to 3 hours, the collected mass was weighed, and the flux was calculated from the mass/time data. The water content in the permeate was analysed using gas chromatography or refractive index measurements.

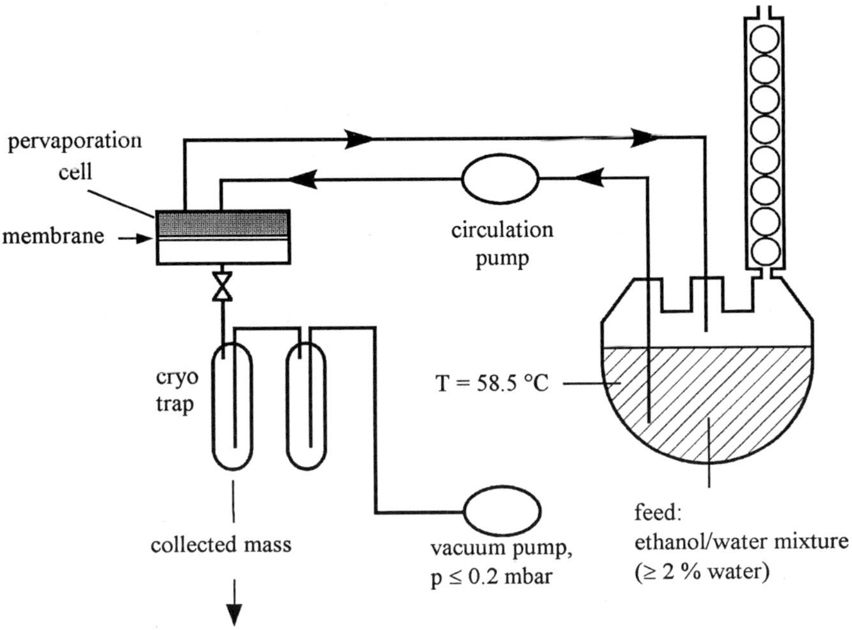


- weight determination → flux J [g $m^{-2}$ $h^{-1}$]
- GC analysis → separation factor $\alpha = \dfrac{[H_2O]_{perm} / [Ethanol]_{perm}}{[H_2O]_{feed} / [Ethanol]_{feed}}$

Fig. 2: Schematic diagram of the pervaporation apparatus.

In Fig. 3, flux and water content of the permeate are plotted as a function of the number of adsorbed layer pairs of polyallylamine/polystyrenesulfonate (PAH/PSS)[9]. Three main results were found: Firstly, the water content in the permeate was always higher than in the feed solution, which only contained 6,2 %(w/w). This indicates preferential transport of water across the membrane. Secondly, an increase of the membrane thickness led to a decrease of the flux and an increase of the water content in the permeate. Thirdly, annealing of the membrane prior to pervaporation led to a decrease of the flux and an improvement of the separation capability. Similar effects are known from other membranes and were discussed already previously[15]. However, even after annealing the water content in the permeate was not very high and did not exceed 70 %, the separation

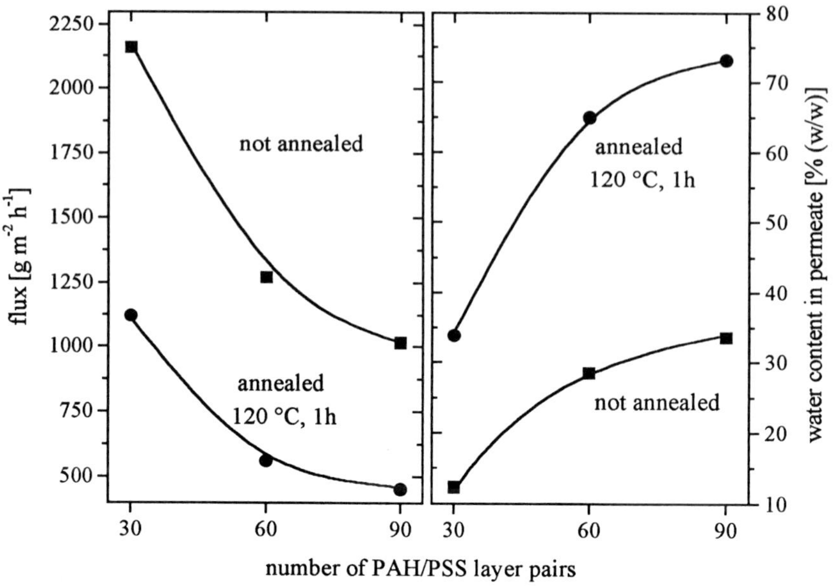


Fig. 3: Effect of the number of adsorbed layer pairs of PAH/PSS on flux and water content in the permeate for samples with and without thermal treatment at 120 °C for 1h. Feed solution: Ethanol/water mixture with 6.2 % (w/w) water[9].

factor $\alpha$ (for definition of $\alpha$ see Fig. 2) was not higher than 70. Obviously, the water-enrichment of the PAH/PSS membrane was only poor.

We therefore tried to improve the separation capability by varying the polyelectrolyte stucture and by optimising the processing parameters. Let us first concentrate on the effect of the polyelectrolyte structure. In Fig. 4, a very idealised model of the polyelectrolyte multilayer membrane is shown. According to this model, the membrane exhibits a physical network structure, the ion pairs representing the cross-linking units. From this model one can derive that cross-linking density and mash size are controlled by the charge density $\rho_c$ of the polyelectrolyte complex. $\rho_c$ can be expressed as the number ratio of ion pairs (= 1) and carbon atoms per repeat unit of the polyelectrolyte complex.

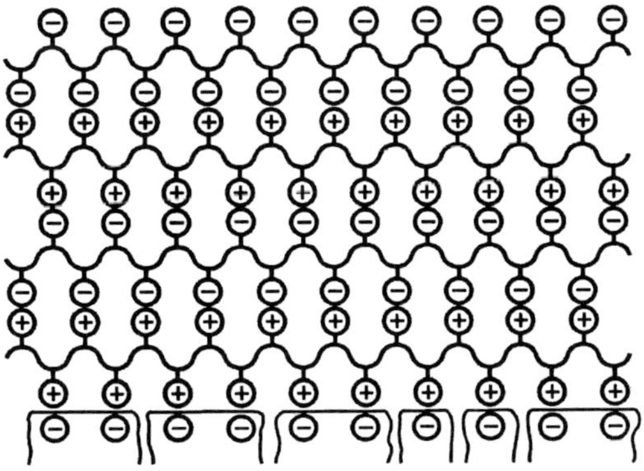

Fig. 4: Idealised structure of the polyelectrolyte complex membrane with the polyelectrolytes being completely ionised.

In Fig. 5, flux and separation behaviour of various polyelectrolyte multilayer membranes are plotted versus the $\rho_C$ parameter. The flux was found to be inversely proportional to $\rho_C$. The water content in the permeate increased with $\rho_C$. Obviously a high separation is obtained at the easiest, if the charge density of the network is high. The clarity of this correlation is surprising, especially because in the definition of $\rho_C$ any contribution from specific structural elements such as the flexibility of the main chain and the side groups, presence of aromatic or aliphatic units, and different polar groups in the polyelectrolytes has been neglected. The origin must be that at a high $\rho_C$ value of the polyelectrolytes only small, highly polar mashes are formed, which are well permeable for the highly polar water molecules, but less permeable for the ethanol molecules because of their lower hydrophilicity.

Besides the chemical structure of the polyelectrolyte, the degree of ionisation of the polar groups is another parameter affecting the charge density of the network. The degree of ionisation is determined by the pH of the polyelectrolyte solution used for membrane preparation. For primary amino groups, for example, a complete protonation is only observed at low pH values. At moderate pH, only partial protonation occurs so that considerably more polyelectrolyte material has to be adsorbed to neutralise the surface charges of the membrane. As a consequence, the thickness of the polyelectrolyte multi-

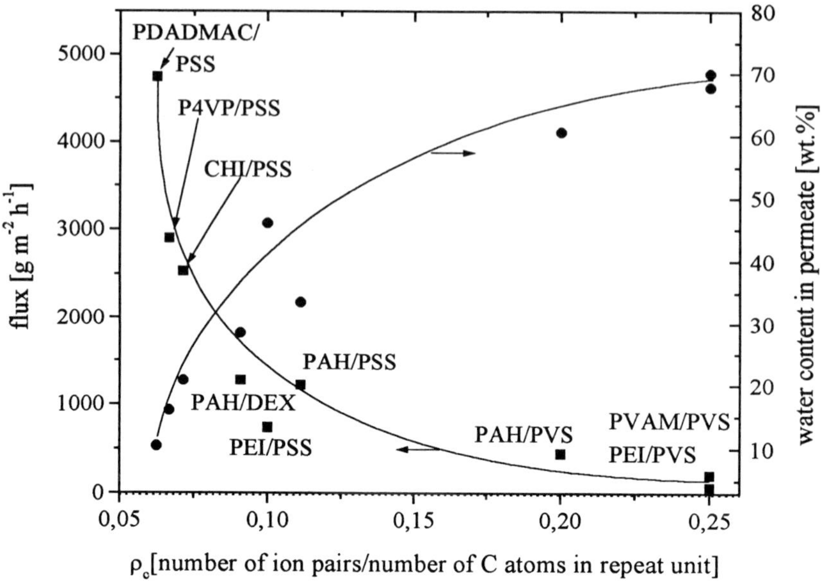


Fig. 5: Dependence of flux and water content in permeate on the charge density $\rho_c$ of the polyelectrolyte complex. Feed solution: 6.2 % (w/w) water; separating membranes always consist of 60 layer pairs of polyelectrolyte complex[12)]. PAH: poly(allylamine); P4VP: poly(4-vinylpyridine); CHI: chitosan; PVAM: poly(vinylamine); PDADMAC: poly(diallyldimethylammonium bromide); PEI: branched poly(ethyleneimine); PSS: poly(styrenesulfonate); DEX: dextrane; PVS: poly(vinylsulfate).

layer is increased[7,16)], while the charge density is decreased. In Tab. 1, flux and water content in the permeate are listed for PAH/PSS membranes prepared at different pH values. As shown in Tab. 1, a maximum water content in the permeate and a minimum in the flux occurs at pH 2,1. At that pH, most of the polar groups of PAH and PSS are present in an ionised state so that upon the salt formation a dense network structure is formed. At higher pH, the amino groups are increasingly deprotonated and the charge density of the membrane is decreased. As a consequence, the flux is increased, and the separation gets worse. As recently shown, the most suitable pH for preparation of a dense network exhibiting maximum separation corresponds with the mean of the $pK_a$ values of the two oppositely charged polyelectrolytes[14)].

Tab.1: Flux and water content in the permeate for PAH/PSS membranes (thickness: 60 layer pairs) prepared from polyelectrolyte solutions of different pH. Feed solution: 6.2% water (w/w).

| pH | $H^+$-conc. [mol $l^{-1}$] | Flux [g $m^{-2}$ $h^{-1}$] | Water content in permeate [%] |
|---|---|---|---|
| 2.44 | 0.00362 | 3860 | 13.0 |
| 2.10 | 0.00794 | 822 | 38.2 |
| 1.79 | 0.0162 | 1270 | 28.5 |

Our study indicates that the ethanol/water separation can be substantially improved, if polyelectrolytes of high charge density are used and if the separating membrane is processed at an optimum pH value of the polyelectrolyte solution. In fact, the plots of Fig. 6a and b indicate that the separation capability of poly(vinylamine)/poly(vinylsulfate) (PVAM/PVS) and poly(ethyleneimine)/PVS (PEI/PVS) separating membranes is superior to the previously described PAH/PSS membrane[10)]. The PVAM/PVS and PEI/PVS membranes are considerably more stable than the PAH/PSS membrane, which is already decomposed at a water concentration in the feed solution above 20 %. The PVAM/PVS membrane is able to enrich water at any composition of the feed solution. For example, at 20 % water in the feed, a separation factor of 700 is reached, while the flux is about 500 g $m^{-2}$ $h^{-1}$, a fairly good value. At higher water content in the feed, still higher flux values are found. A similar behaviour is found for the PEI/PVS membrane except that this membrane is only suited for ethanol/water separation, if the water content in the feed is lower than 50 %.

In Fig. 7a and b, pervaporation of various alcohol/water mixtures through the PVAM/-PVS membrane is characterised. It is shown that mixtures of 1-propanol/water or 2-propanol/water are more efficiently separated than the ethanol/water mixtures, because the transport of the more hydrophobic propanol is less favoured by the highly polar membrane. Except for very small water concentration in the feed, i.e. less than 5 %, the permeate always contains more than 95 % water. For example, the separation factor of the 80/20 mixture of 2-propanol/water is about 7000. Simultaneously, the flux of propanol/water mixtures is much higher than of ethanol/water mixtures, probably because

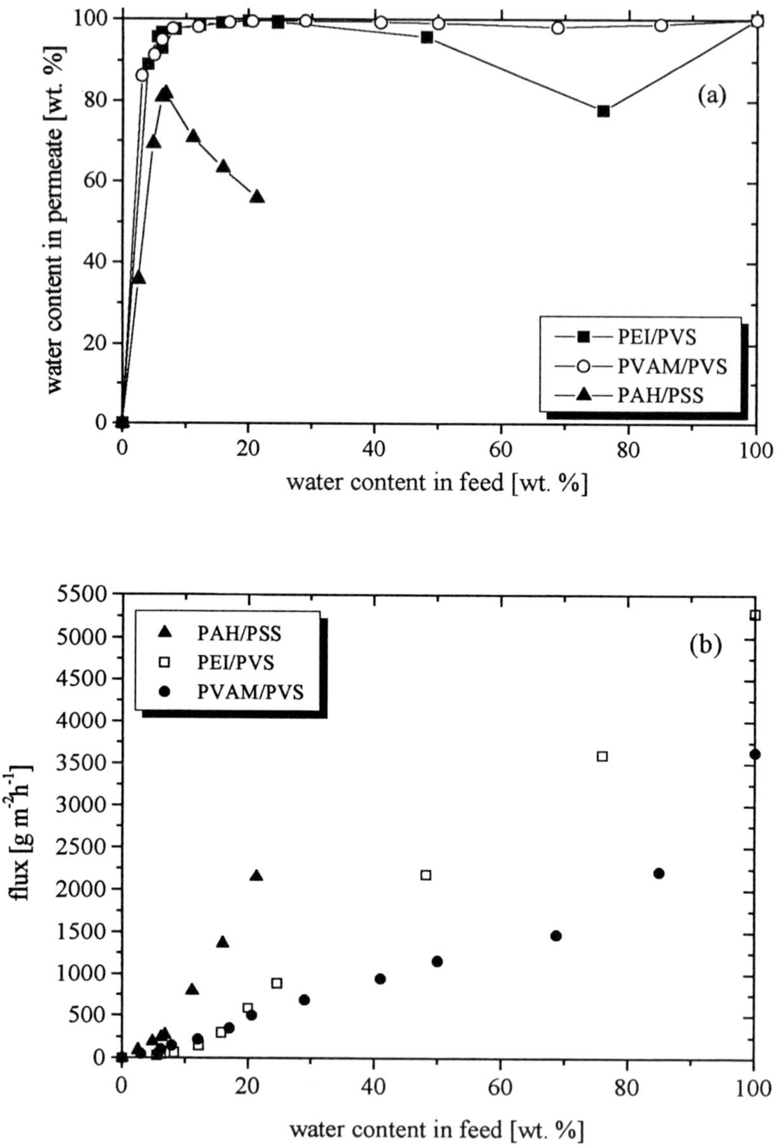


Fig. 6: Dependence of water content in permeate (a) and flux (b) on water content in feed solution for PVAM/PVS, PEI/PVS and PAH/PSS membranes. All samples consist of 60 layer pairs. PAH/PSS and PEI/PVS membranes were annealed at 90 °C for 1h prior to pervaporation[12].

of the weaker hydrogen bonding between propanol and water molecules. For example, the 80/20 mixture of 2-propanol/water exhibits a high flux of 1,5 kg $m^{-2}$ $h^{-1}$, while for an ethanol/water mixture of the same composition the flux is only 0,5 kg $m^{-2}$ $h^{-1}$.

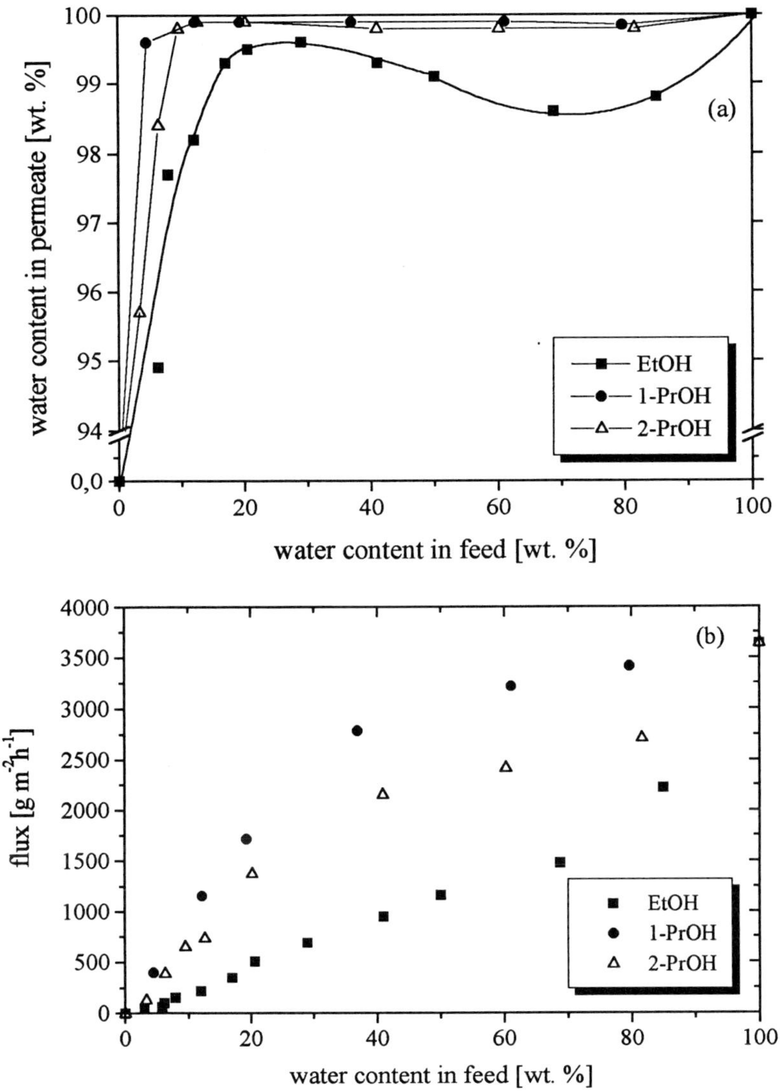


Fig. 7: Dependence of water content in permeate (a) and flux (b) on water content in feed solution for various alcohol/water mixtures. Separating membrane: 60 layer pairs of PVAM/PVS.

## Ion separation

The ion permeation through solution-cast bipolar membranes consisting of both negatively and positively charged polyelectrolyte layers was already investigated several years ago[17,18]. It was found that divalent cations receive a much stronger repulsive force from positively charged layers than monovalent ones, and that the same holds for the divalent anions, which are rejected by the negatively charged layers. Consequently, the bipolar membranes represent barriers for divalent ions in general and can be used for ion separation[7,18]. The self-assembled polyelectrolyte multilayers consist of cationic and anionic polyelectrolytes in alternating fashion and therefore can be described as a multi-bipolar membrane on a molecular level, which likewise should be able to separate mono- and divalent ions. In Fig. 8, a model of the ion rejection by the self-assembled multilayer membrane is represented. The model implies that the ion separation becomes progressively more effective, if the number of adsorbed polyelectrolyte layers is increased.

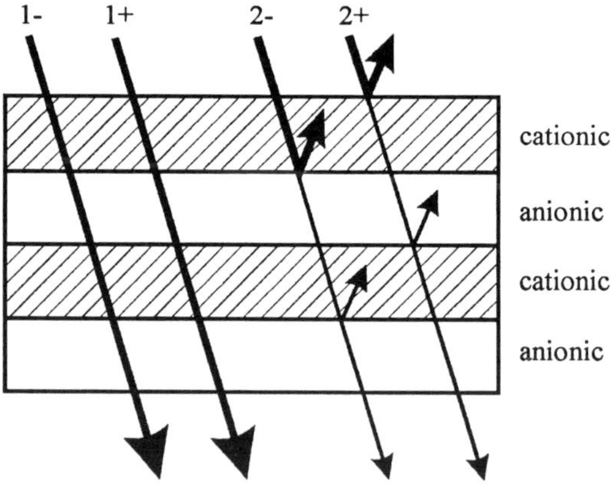


Fig. 8: Rejection model of a multi-bipolar membrane[9].

In order to study the ion permeation, the cell shown in Fig. 9 was used. The cell consists of two chambers separated by the membrane. One of the chambers contains the aqueous salt solution (concentration: 0.1 mol $l^{-1}$) and the other one pure water. From the measurement of the initial increase in conductivity ($\Delta\Lambda/\Delta t$) in the water chamber, the permeation rate $P_R$ was calculated using the equation shown in Fig. 9. A possible contribution to ($\Delta\Lambda/\Delta t$) originating from the osmotic flow of water in the opposite direction was not corrected.

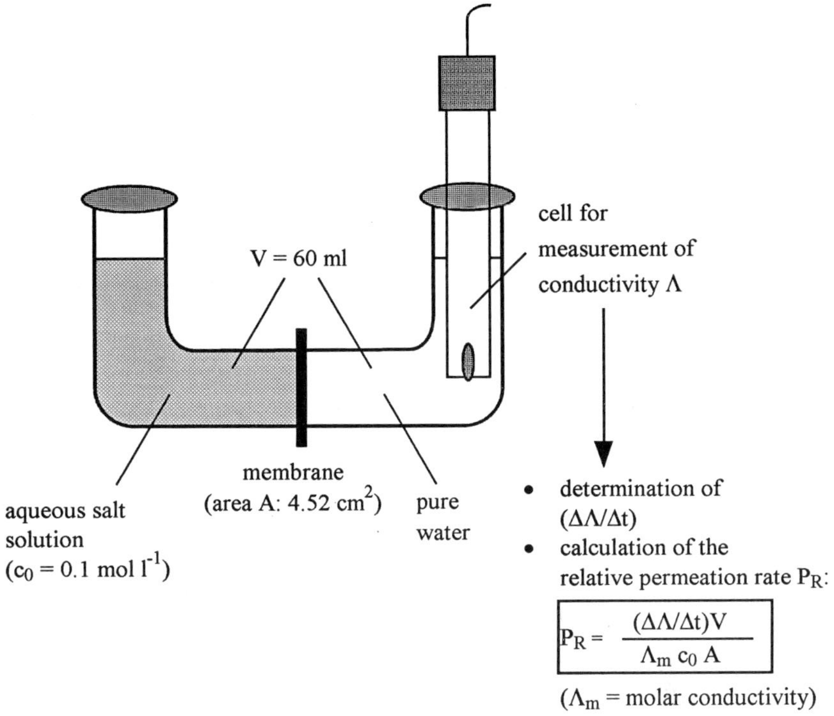


Fig. 9: Schematic representation of the cell used for measurement of ion permeation.

In Fig. 10, the permeation rates of NaCl and $MgCl_2$ are plotted versus the thickness of the membrane. With increasing thickness, the permeation rates of the two salts are decreased, but for $MgCl_2$ the decrease is much stronger than for NaCl. Therefore, the separation factor $\alpha$ (for definition of $\alpha$ see Fig. 9) strongly increases with the mem brane thickness. For example, a sample containing 60 layer pairs of PAH/PSS exhibits separation factors $\alpha$ of 15,1 for NaCl/$MgCl_2$, and of 9,9 for NaCl/$Na_2SO_4$.

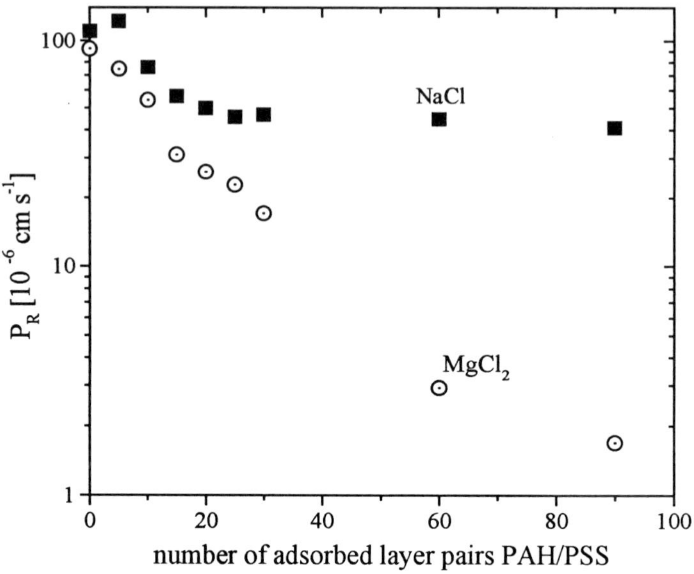


Fig. 10: Effect of the thickness of the separating membrane (PAH/PSS) on permeation rates $P_R$ of NaCl and $MgCl_2$ [8].

The ion permeation is strongly affected by the structure of the polyelectrolytes used for membrane preparation. In Fig. 11, the permeation rates of NaCl and $MgCl_2$ are plotted as a function of the charge density $\rho_c$ of the polyelectrolyte membrane. The higher the charge density, the lower is the permeation rate, and the difference between the salts with mono- and divalent ions becomes progressively larger with increasing $\rho_c$. The plots also indicate that membranes with PAH as the cationic polyelectrolyte always exhibit lower permeation rates than membranes containing other polycations. The origin for the especial behaviour of PAH is not yet clear.

Besides the polyelectrolyte structure, the processing parameters such as salt content and pH of the polyelectrolyte solution have a strong influence on the ion permeation, too. The pH controls the ionisation of the polyelectrolytes, which in its turn influences the charge density of the membrane, as discussed above. As recently shown, a low pH value

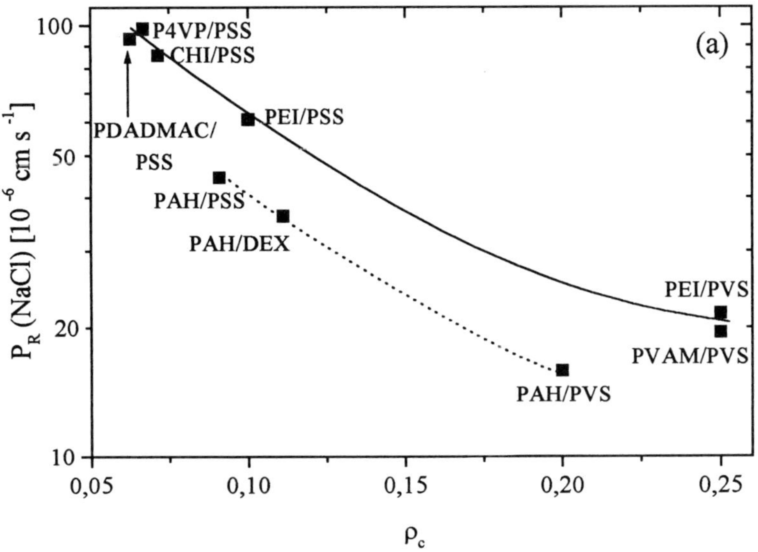


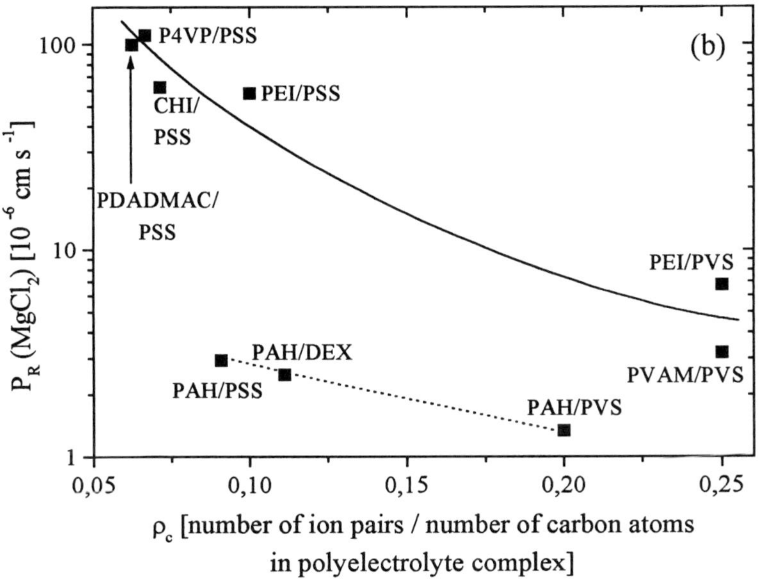


Fig. 11: Effect of charge density $\rho_c$ of polyelectrolyte multilayer on permeation rate $P_R$ of (a) NaCl and (b) $MgCl_2$. Separating membranes always consist of 60 layer pairs of polycation polyanion.

favours the formation of a dense, less permeable membrane of high charge density, which exhibits an improved rejection of divalent ions. Upon salt addition to the polyelectrolyte solution, the coil dimensions of the polyelectrolytes are reduced. As a consequence, the polyelectrolytes are no longer adsorbed in a flat conformation, but as coils, so that the individual layers are much thicker[7] and the permeation rates are decreased. Since especially the $P_R$ values of the salts with divalent ions are decreased, considerably higher separation factors are obtained (Tab. 2).

Tab. 2: Permeation rates $P_R$ of NaCl, $MgCl_2$ and $Na_2SO_4$ and selectivities $\alpha$ for PAH/PSS and PVAM/PVS membranes prepared from salt-free and salt-containing polyelectrolyte solution (thickness: 60 layer pairs).

| Membrane | addition of salt | $P_R$ (NaCl) [$10^{-6}$ cm $s^{-1}$] | $P_R$ ($MgCl_2$) [$10^{-6}$ cm $s^{-1}$] | $P_R$ ($Na_2SO_4$) [$10^{-6}$ cm $s^{-1}$] | $\alpha$ (NaCl/ $MgCl_2$) | $\alpha$ (NaCl/ $Na_2SO_4$) |
|---|---|---|---|---|---|---|
| PAH/ PSS | + | 22.5 | 0.2 | 0.5 | 112.5 | 45.0 |
| PAH/ PSS | - | 44.4 | 2.9 | 4.5 | 15.1 | 9.9 |
| PVAM/ PVS | + | 3.1 | 0.06 | 0.3 | 51.7 | 10.3 |
| PVAM/ PVS | - | 19.5 | 2.4 | 11.5 | 8.1 | 1.7 |

## Summary and Conclusions

Our study indicates that alternating electrostatic adsorption of cationic and anionic polyelectrolytes on porous substrates is a versatile method to prepare composite membranes with ultrathin, pore-free separation layer. By careful choice of the polyelectrolytes and by optimisation of the processing conditions, membranes with excellent separation capability for alcohol/water mixtures or mono- and divalent ions can be tailored. It is also shown that the transport of small molecules and ions through the membrane is strongly a function of the density of the network formed upon the complex formation between the oppositely charged polyelectrolytes. For the separation processes discussed above, polyelectrolytes with high charge density are especially suited. They are able to form a close, highly polar network structure, which favours the transport of highly polar liquids such as water, and retards the transport of more hydrophobic alcohols, and which rejects the divalent ions of high charge density more strongly than the monovalent ions. In further experiments, the separation of organic liquids of different hydrophilicity will be studied, as well as the ion permeation under reverse osmosis conditions.

**Acknowledgements.** The authors are grateful to Dr. H. Scholz, Sulzer Chemtech, Neunkirchen Dr. G. Ellinghorst, FhG Bremen, for providing the PAN/PET membranes and for helpful discussions, and to Dr. Mahr (BASF, Ludwigshafen) for providing polyvinylamine. Financial support from the Deutsche Forschungsgemeinschaft (Ti 219/3-3) is also gratefully acknowledged.

## References

1 K. Scott; *Handbook of Industrial Membranes*, 2nd edition, Elsevier Adv. Technol., Oxford (1998)
2 W. Pusch, A. Walch; *Angew. Chem.* Int. Ed. Engl. **21**, 660 (1982)
3 G. Decher; *Science* **277**, 1232 (1997)
4 P. Stroeve, V. Vasquez, M.A.N. Coelho, J.F. Rabolt; *Thin Solid Films* **284-285**, 708 (1996)
5 J.-M. Leväsalmi, T. J. Mc Carthy; *Macromolecules* **30**, 1752 (1997)
6 F. v. Ackern, L. Krasemann, B. Tieke; *Thin Solid Films* **329,** 762 (1998)
7 G. Decher, J. Schmitt; *Prog. Colloid Polym. Sci.* **89**, 160 (1992)
8 L. Krasemann, B. Tieke; *Langmuir* **16**, 287 (2000)
9 L. Krasemann, B. Tieke; *Mater. Sci. Eng.* C **8-9**, 513 (1999)
10 L. Krasemann, B. Tieke; *J. Membr. Sci.* **150**, 23 (1998)
11 L. Krasemann, B. Tieke; *Chem. Eng. Technol.* **23**, 211 (2000)
12 L. Krasemann, A. Toutianoush, B. Tieke; *J. Membr. Sci.* 2000, submitted
13 J. Meier-Haack, Th. Rieser, W. Lenk, D. Lehmann, S. Berwald, S. Schwarz; *Chem. Ing. Tech.* **71,** 839 (1999)
14 J. Neel, *Introduction to Pervaporation*, in R.Y.M. Huang (ed.), *Pervaporation Membrane Separation Processes*, Elsevier, Amsterdam (1991), p.42
15 W. Pusch, A. Delyannis, E.E. Delyannis; *Desalination* **16,** 65 (1975)
16 D. Yoo, S.S. Shiratori, M.F. Rubner; *Macromolecules* **31**, 4309 (1998)
17 M. Urairi, T. Tsuru, S. Nakao, S. Kimura; *J. Membr. Sci.* **70**, 153 (1992)
18 T. Tsuru, S. Nakao, S. Kimura; *J. Membr. Sci.* **108**, 269 (1995)

# Tailoring of polymer properties in segmented block copolymers

Doris Pospiech*, Liane Häußler, Kathrin Eckstein, Dieter Voigt, Dieter Jehnichen, Antje Gottwald, Wolfram Kollig, Andreas Janke, Karina Grundke, Carsten Werner, Hans R. Kricheldorf[#]

Institute of Polymer Research Dresden, Hohe Str. 6, D-01069 Dresden, Germany

[#] Institute of Technical and Macromolecular Chemistry, University of Hamburg, Bundesstr. 45, D-20146 Hamburg, Germany

SUMMARY: Linking segments of two different polymers in block copolymers offers the opportunity to combine their macroscopic properties and to design tailor-made materials. This is shown for segmented block copolymers $(A\text{-}B)_n$ in which the basic segment is polysulfone. Incorporation of either hydrophilic and or hydrophobic segments results in a dramatic change of surface properties which is also correlated to the bulk structure of the block copolymers.

## Introduction

Block copolymers (BCP) have been intensively studied for more than 30 years. The intention was always to combine properties of two or more polymers in a new polymer, resulting in materials with tailor-made property profiles. Thus, a variety of block copolymers with different structure was reported and the phase separation between the block copolymer segments was shown to cause property combinations and synergisms [1-3]. Properties to be adjusted can be processing behavior (melt and solution viscosity), mechanical and thermal performance and also surface properties (e.g., to achieve biocompatibility)[4]. Tools to design new materials by block copolymer formation are to choose the polymeric segments to be incorporated with respect to their macroscopic properties, i.e., to select which properties should be developed in the new material, and, on the other hand, the knowledge of the phase separation behavior because the properties of block copolymers are mainly controlled by the microphase-separated morphology. The conditions for phase separation are relatively well-known for certain diblock copolymers (e.g., polystyrene-b-polyisoprene) expressed in phase diagrams that were obtained both experimentally and by calculations[5-7].

 CCC 1022-1360/00/$ 17.50+.50/0

We have been investigating segmented block copolymers of the structure $(A\text{-}B)_n$ having polysulfone (PSU) blocks as building blocks in order to introduce the advantageous properties of PSU, e.g., excellent mechanical strength, temperature- and oxydation stability and chemical resistance, into block copolymers. It could be shown that incorporation of liquid-crystalline polyester and poly(ester imide) segments actually resulted in a combination of mechanical properties in phase separated samples, whereas non-phase separated BCP developed property synergisms particularly with respect to the glass transition temperature[8-10]. The processability in the melt as well as the thermostability could generally be improved by block copolymer formation in comparison to the homopolymers.

In this work we will discuss the influence of phase separation in segmented block copolymers with surface active segments. Both, hydrophilic poly(tetramethylene oxide) segments and semifluorinated polyester segments having low surface energy were built into the polysulfone chain. The investigations were performed to address the questions whether it is possible to understand the phase separation behavior of these rather complicated structures and, on the other hand, to investigate whether or not the degree of phase separation in the bulk influences the surface properties of the polymers.

## Synthesis of the segmented block copolymers

The segmented block copolymers were synthesized by a transesterification poly-condensation process in the melt, as it was described for several LC block copolymers in our earlier work[8-11]. Thus, acetoxy-terminated PSU oligomers with molecular weights ranging from 1000 to 10000 g/mol were reacted with carboxy-terminated oligomers, as schematically shown in Equation 1.

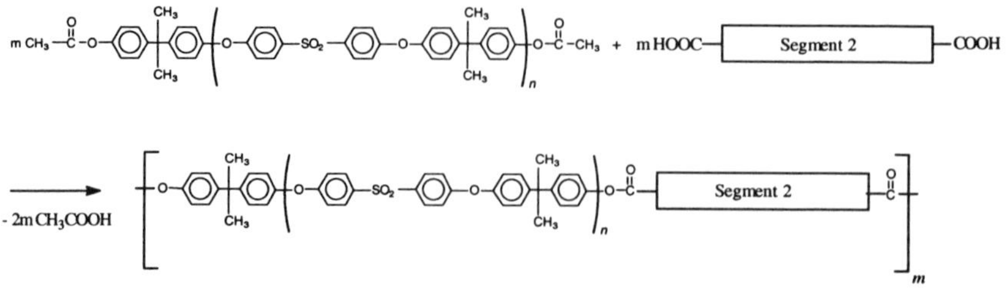


**Equation 1**

Relative molecular weights were calculated using PS calibration. The GPC molecular weights obtained for different types of segmented block copolymers are given in Tab. 1.

Table 1. Molecular weights of different types of segmented block copolymers.

| **Type of segmented block copolymer** | **$M_w$ (GPC) g/mol** | **$M_w/M_n$** |
|---|---|---|
| Polysulfone-Poly(tetramethylene oxide) **1** | 18-45,000 [a)] | 1.8-2.9 |
| Polysulfone-(Semifluorinated polyester) **2** | 20-40,000 [b)] | 1.7-2.7 |
| Polysulfone-Poly(ethylene terephthalate-co-oxybenzoate) **3** | 100,000 [c)] | 1.7-3.9 |
| Polysulfone-Poly(ethylene terephthalate-co-polyester imide) **4** | 50,000 [c)] | 2.1-2.6 |
| Polysulfone-Poly(ester imide) **5** | 45-70,000 [c)] | 2.3-5.2 |

a) GPC: Knauer; eluent: $CHCl_3$; separation column: Merck LiChrogel PS 40; RI detection
b) GPC: Knauer; eluent: pentafluorophenol/$CHCl_3$ (1/3 vol/vol); separation column: PSS-SDV linear; RI detection
c) GPC: Knauer; eluent: cresol/$CHCl_3$ (1/3 vol/vol); separation column: Waters Styragel HT3/HT6E; RI detection

For the results presented here, two different series of segmented block copolymers **1** and **2** (chemical structure as given in Formulae **1** and **2**) have been investigated. The chemical characterization of the samples is given in Tab. 2.

**Polysulfone - Poly(tetramethylene oxide) (PSU-PTMO) block copolymers**

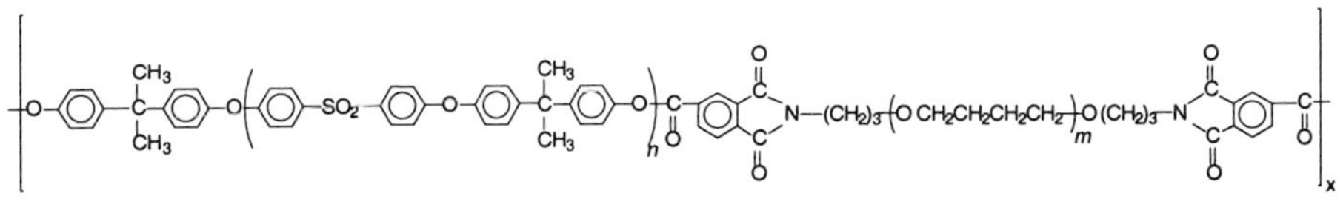

**1**

**Polysulfone - semifluorinated polyester (PSU-SF-PES) block copolymers**

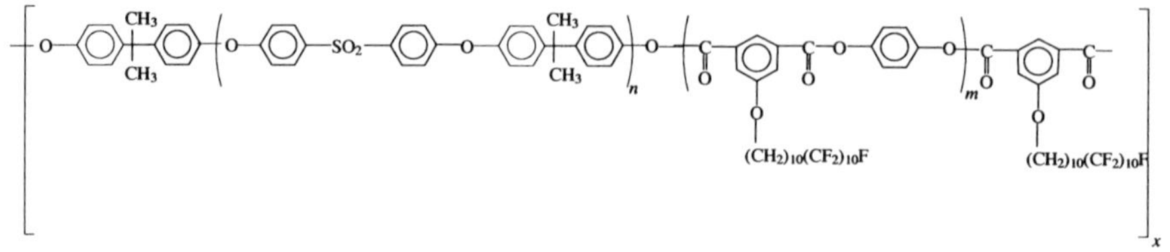


**2**

Table 2. Characterization of the segmented block copolymers under discussion.

| Block copolymer | PSU segment $M_n$ g/mol | Second segment $M_n$ g/mol | BCP $M_w$ (GPC) g/mol a) | BCP $M_w/M_n$ |
|---|---|---|---|---|
| | | **PTMO** | | |
| **1**-1a | 1150 | 283 | 32,300 | 3.80 |
| **1**-1b | 1150 | 1150 | 17,700 | 2.83 |
| **1**-1c | 1150 | 2500 | 18,400 | 2.49 |
| **1**-1d | 1150 | 3600 | 25,600 | 3.44 |
| **1**-1e | 1150 | 5200 | 45,200 | 1.75 |
| **1**-1f | 1150 | 9100 | 35,800 | 2.07 |
| **1**-2a | 2500 | 283 | 43,500 | 2.54 |
| **1**-2b | 2500 | 1150 | 22,300 | 2.93 |
| **1**-2c | 2500 | 2500 | 23,000 | 2.38 |
| **1**-2d | 2500 | 3600 | 19,300 | 2.12 |
| **1**-2e | 2500 | 5200 | 26,600 | 1.80 |
| **1**-2f | 2500 | 9100 | 43,300 | 1.82 |
| | | **SF-PES** | | |
| **2**-1a | 2300 | 840 | 10,600 | 2.36 |
| **2**-1b | 2300 | 1750 | 29,000 | 1.82 |
| **2**-1c | 2300 | 3600 | 20,100 | 2.63 |
| **2**-1d | 2300 | 9000 | 21,000 | 2.13 |
| **2**-2a | 9000 | 840 | 44,000 | 2.16 |
| **2**-2b | 9000 | 1750 | 30,400 | 2.64 |
| **2**-2c | 9000 | 9000 | 27,400 | 1.68 |

a) GPC: Knauer; eluent: $CHCl_3$; separation column: Merck LiChrogel PS 40; RI detection

b) GPC: Knauer; eluent: pentafluorophenol/$CHCl_3$ (1/3 vol/vol); separation column: PSS-SDV linear; RI detection

## Incorporation of hydrophilic poly(tetramethylene oxide) segments

The synthesis and characterization of segmented PSU/PTMO block copolymers was reported in detail in a previous paper[12]. The block copolymers of series **1**-1 with short PTMO segments (1150 g/mol) have a non-crystalline, homogeneous bulk structure, showing a single $T_g$ according to the miscibility rule. The block copolymers of series **1**-2 containing PTMO segments with a molecular weight of 2500 g/mol are separated into two amorphous phases. Samples with low molecular weight polysulfone segments contain additionally crystallized PTMO. However, the amorphous phases, although phase-separated, are significantly mixed resulting in a strong decrease of the $T_g$ of the PSU phase. The phase separation behavior of the PSU/PTMO block copolymers observed experimentally agrees well with phase diagrams obtained by Mean Field calculations for polydisperse segmented block copolymers[13].

Introduction of hydrophilic PTMO units into the PSU chain actually leads to a significant hydrophilization of polysulfone. The dynamic contact angles (advancing and receding contact angles) measured on thin films prepared by spin coating increase continuously in the first, non-phase separated series (Fig.1).

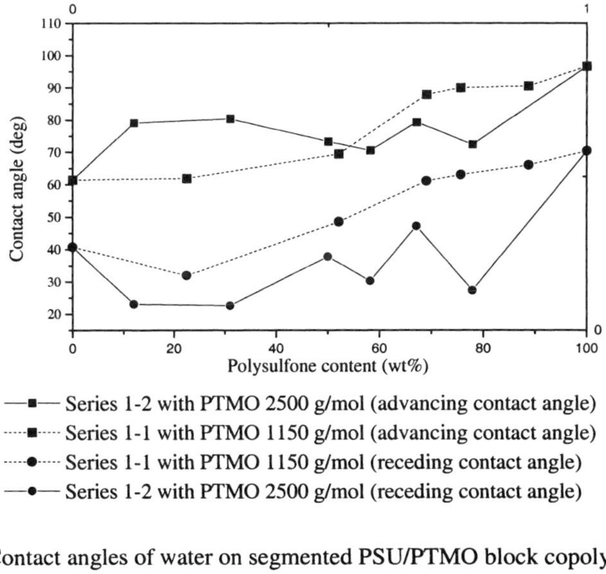


Fig. 1: Contact angles of water on segmented PSU/PTMO block copolymer films

By XPS investigation an enrichment of the surface with PTMO units was observed. The depth profiles show a continuous drop of the S/C ratio with decreasing sampling depth. The phase separated block copolymers (series **1**-2) show a discontinuous increase in contact angle with increasing PSU content of the block copolymers. The contact angle hysteresis is much higher than in homogeneous BCP thus indicating that the phase separation in the bulk also produces chemically heterogeneous surfaces having both, hydrophilic as well as hydrophobic parts at the surface. The XPS depth profiles (S/C ratio, indicating the relative concentration of PSU in different sampling depths) depend on the BCP segment molecular weights. The S/C ratio increases, as in series **1**-1, in samples with short PSU segments (enrichment with PTMO) whereas samples with long PSU segments show higher relative concentrations of PSU at the surface than in greater depth (enrichment with PSU).

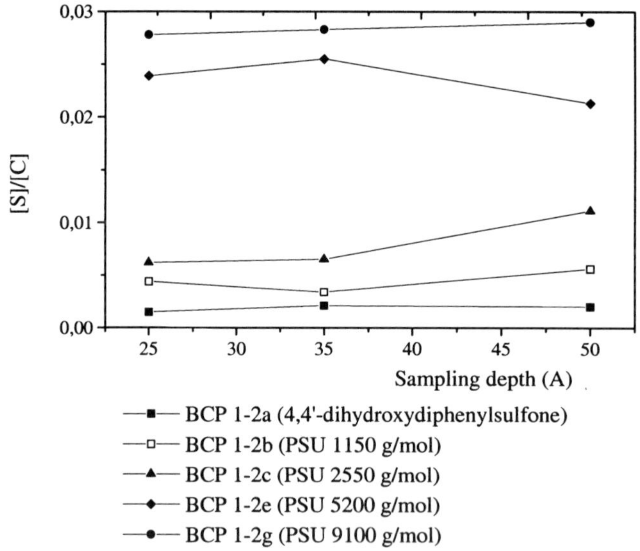


Fig. 2: XPS analysis of the surface composition of block copolymers **1**-2 (depth profile).

This is probably caused by the presence of PTMO crystallites near the surface (Fig. 3, left). In contrast, the PSU segments as low energy component are able to segregate to the surface in BCP with long PSU spacers producing a more uniform surface morphology (Fig. 3, right).

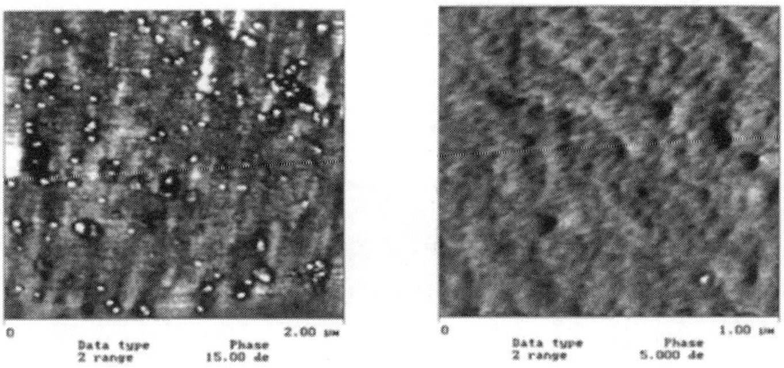


Fig. 3: AFM micrographs of surfaces of BCP 1-2c (left) and BCP 1-2g (right), phase contrast by stiffness; bright: hard.

A next question was whether the hydrophilic properties of PTMO/PSU BCP could be retained in physical mixtures with polysulfone. Thus, 50/50 wt/wt blends of PSU and block copolymers of series **1** were prepared by solution casting. DSC analysis revealed that different states of miscibility (demixed to completely homogeneous) can be achieved by just selecting the appropriate molecular weight of polysulfone segments in the BCP[14)]. The wetting behavior of thin films on Si wafer was investigated by contact angle measurements (Fig. 4). The contact angles obtained for the blends are significantly lower than that of pure polysulfone, i.e., the hydrophilic properties of the block copolymers determine the surface properties of the blend. The state of miscibility is reflected by the contact angle hysteresis (i.e., the difference between advancing and receding dynamic contact angle). The hysteresis is higher for completely immiscible blends (~ 45 deg for BCP **1**-1a), indicating chemically heterogeneous surfaces, whereas on partially miscible and miscible blends the contact angle hysteresis typically obtained is between 25 and 30 deg. This result offers also a tool to create chemically heterogeneous surfaces.

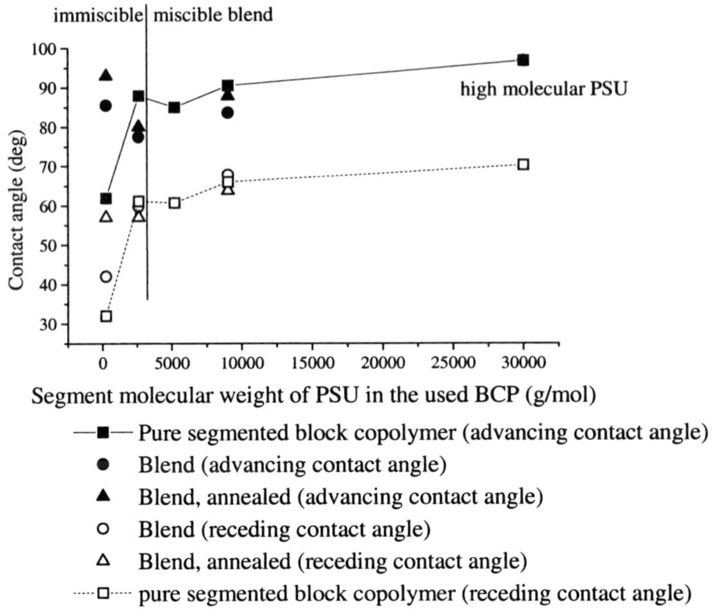


Fig. 4: Advancing and receding contact angles using water as probe obtained on thin films of blends from PSU and PTMO 1150/PSU segmented block copolymers

## Incorporation of low surface energy segments (SF-PES)

Poly(p-phenylene-1-oxydecylperfluorodecyl-isophthalate) has been synthesized to introduce semifluorinated side chains to a polyester with fully aromatic backbone[15]. Due to microphase separation between the alkyl- and the perfluoroalkyl part of the side chain, the polymer forms a self-organizing, layered solid phase structure and thermotropic liquid-crystalline behavior. The highly ordered bulk structure results also in a highly ordered surface structure having $CF_3$-groups at the surface which consequently leads to an extremely low surface energy ($\gamma_{sv}$ = 9.8 mN/m). Linking such segments to polysulfone units in block copolymers should change the surface properties of polysulfone dramatically, though retaining the excellent film-forming properties of PSU. The huge Flory-Huggins interaction parameter calculated for the block copolymer ($\chi$ = 13.6) indicates that phase separation in these BCP would occur at very low segment molecular weights. The onset of phase separation was examined by DSC. Fig. 5 displays the DSC curves for PSU-SF-PES BCP of series **2**-1.

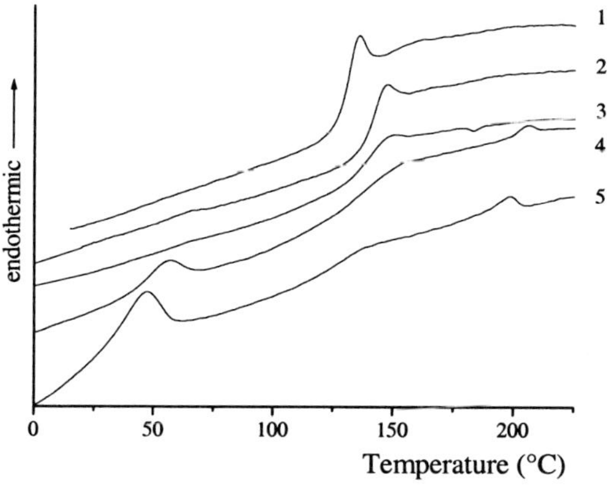


Fig. 5: DSC curves (2nd heating) of PSU-SF-PES block copolymers (series **2**-1).

1: Polysulfone (2300 g/mol)
2: BCP **2**-1b: SF 1750/ PSU 2300
3: BCP **2**-1c: SF 3600/ PSU 2300 BCP **2**-1c: SF 3600/ PSU 2300
4: BCP **2**-1d: SF 9000/ PSU 2300
5: Semifluorinated polyester

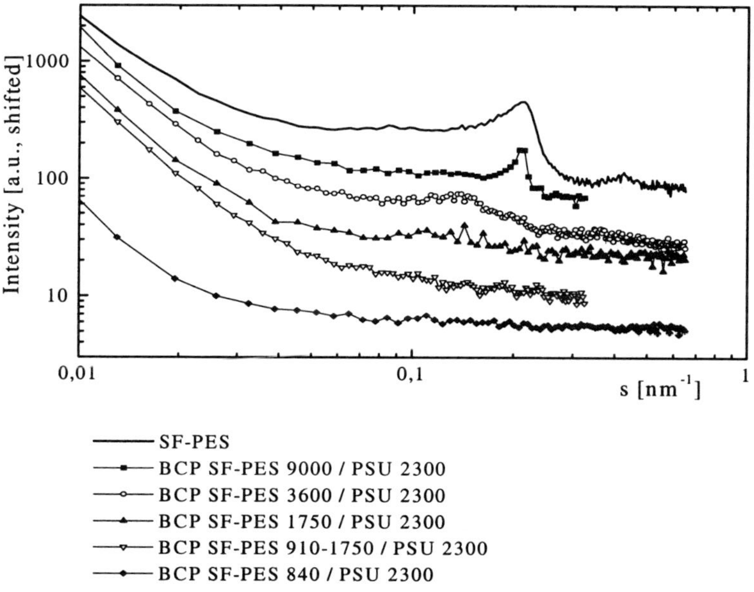


Fig. 6: SAXS curves of polysulfone-semifluorinated polyester block copolymers

After the state of phase separation in the PSU/SF-PES block copolymers was unambigously ascertained, the surface properties of thin film were characterized by XPS, contact angle measurements and SFM. The XPS depth profiles of films after annealing for 4 h at 200°C, i.e., 20 K above the $T_g$ of the PSU phase, reveal a significant enrichment of the surface with fluorine, as seen by the ratio of F/C (Fig. 7).

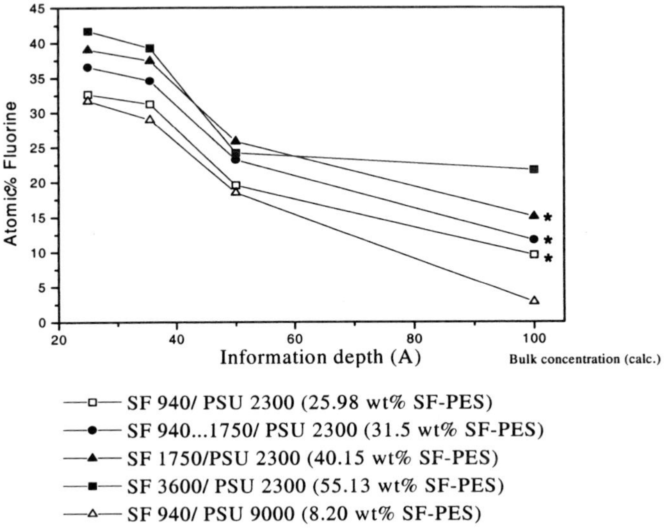


Fig. 7: XPS depth profiles of PS/SF-PES block copolymers
(*: phase separated in the bulk)

Plotting the surface free energies of the block copolymer samples obtained by contact angle measurements (calculation by the Equation-of-State approach[16]) versus the actually measured fluorine contents at the surface (Fig. 8) shows that the bulk morphology obviously does not influence the surface energy, i.e., it depends only on the fluorine concentration at the surface which is determined by the strong tendency of the fluorinated segments to segregate to the surface. The fluorinated segments of the block copolymer form disperse domains at the surface as detected by Scanning Force Microscopy. The size of the domains depends on the molecular weight of the semifluorinated polyester segment, i.e., on the block copolymer composition, as discussed recently[17]. The contact angles raise with increasing bulk concentration of fluorinated segments. The surface free energies of block copolymers are in the range of 12 to 16 mN/m which is much less than that of polysulfone. That means, introduction of hydrophobic segments into polysulfone again leads to a significant alteration of surface properties of the block copolymer.

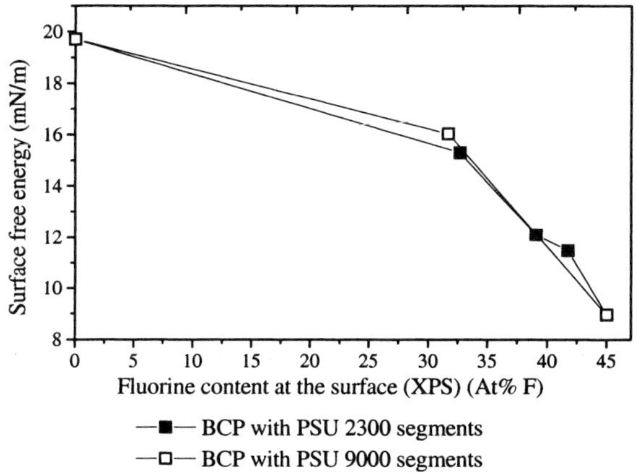


Fig. 8: Surface free energies of PSU/SF-PES block copolymers versus the fluorine content at the surface, obtained by XPS
(0 %F: polysulfone; 45 % F: semifluorinated polyester)

## Adsorption of model proteins

Both, hydrophilic as well as strongly hydrophobic materials can prevent the adsorption of proteins which is a significant point to prevent biofouling, e.g. for membranes. Therefore, the interaction of the modified polysulfone surfaces with two model proteins, Human Serum Albumin (HSA, molecular weight: 68,000 g/mol) and Fibrinogen (Fib, molecular weight: 360,000 g/mol), in phosphate buffer solution was examined by variable angle spectroscopic ellipsometry (VASE). VASE allows to determine time-resolved the thickness of the protein layer, shown in Fig. 9 for a polysulfone film. It can be seen that the thickness of the protein film increases continuously reaching a maximum after certain time. The surface concentration of protein on the polymer film at the maximum was then calculated according to de Fejter[18].

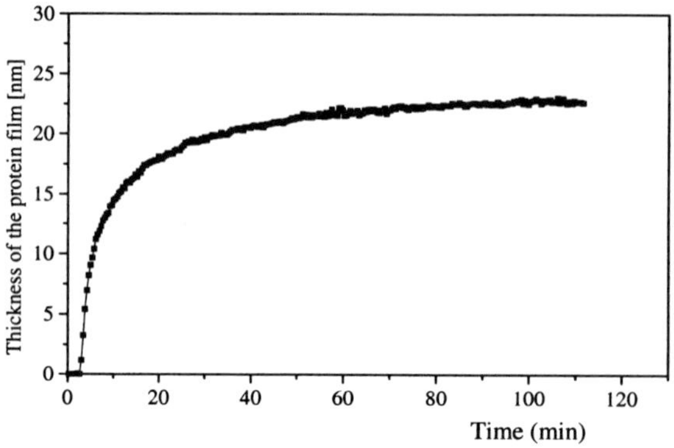


Fig. 9: Time-resolved measurement of the thickness of Fibrinogen adsorption (Fib concentration: 200 ppm in buffer) to a polysulfone film (spincoated, dried in vacuum and annealed for 4 h at 200 °C).

Fig. 10 compares the surface concentration Γ of HSA on different block copolymer films (plotted versus the surface free energy of the BCP).

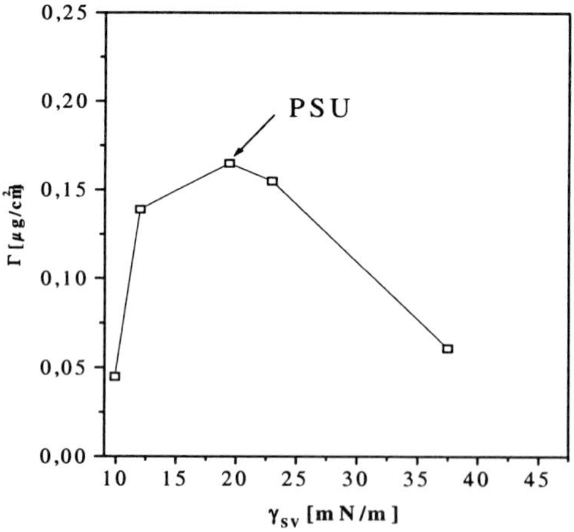


Fig. 10: Surface concentration of Human Serum Albumin (concentration: 100 ppm in buffer) after 150 min measuring time versus the surface free energy of different polysulfone block copolymer films (value at $\gamma_{sv}$ = 19.5 mN/m: polysulfone)

The surface concentration of Human Serum Albumin on the block copolymer films is lower than that on the pure polysulfone film. Both, hydrophilic as well as strongly hydrophobic modification results in a decrease of protein adsorption. The same tendency was found when using the larger Fibrinogen as model protein. Obviously, protein adsorption is significantly prevented by the ultralow surface free energies of the fluorinated samples, whereas on the other hand the low surface tension between the hydrophilic PSU/PTMO block copolymers and the protein in solution has the same effect. Detailed investigation is in progress and the results will be published soon elsewhere.

## Conclusion

It was demonstrated that the synthesis of multiblock copolymers of polysulfone containing either hydrophilic PTMO segments or ultrahydrophobic semifluorinated polyester segments leads to a significant modification of surface properties of polysulfone. The state of phase separation of the block copolymers in the bulk can be controlled by the segment molecular weights of the blocks used. In the case of segmented PSU/PTMO block copolymers phase separation in the bulk produces chemically heterogeneous surfaces. In contrast, block copolymers having semifluorinated polyester segments in the chain show surface segregation of the fluorinated segments independently on the state of phase separation resulting in materials with very low surface free energies. Both approaches finally help to reduce the adsorption of model proteins significantly.

## Acknowledgement

Financial support by *Deutsche Forschungsgemeinschaft* (grant number Po 575/2-2) is gratefully acknowledged. The authors also would like to thank Dr. Hartmut Komber for NMR investigation, Dr. Frank Simon and Mr. Dieter Pleul (IPF Dresden) for XPS measurements. The Silicon wafers used as substrate for thin polymer films were kindly supplied by Wacker.

## References


1. A. Noshay, J. E. McGrath: *Block Copolymers: Overview and Critical Survey,* Academic Press, New York 1977, pp.25 and pp. 305
2. R. W. Richards, in: *Multicomponent Polymeric Systems",* I. S. Miles, S. Rostami (Eds.), Longman Publ., Harlow, UK, 1992
3. I. Goodman (Ed.): *Developments in Block Copolymers",* Appl. Sci. Publ. , New York, Basel, Vol. 1 (1982), Vol 2 (1985)
4. R. Mülhaupt, U. Buchholz, J. Rösch, N. Steinhauser, *Angew. Makromol. Chem.* **223**, 47 (1994)
5. F. S. Bates, *Science* **251**, 898 (1991)
6. F. S. Bates, G. H. Frederickson, *Physics Today* **2**, 32 (1999)
7. G. H. Fredrickson, E. Helfand, F. S. Bates, L. Leibler, *Springer Ser. Chem. Phys.* **51**, 13 (1989)
8. D. Pospiech, L. Häußler, K. Eckstein, H. Komber, D. Voigt, D. Jehnichen, E. Meyer, A. Janke, H. R. Kricheldorf, *Designed Monomers and Polymers* **1**, 2, 187 (1998)
9. D. Pospiech, L. Häußler, B. Voit, F. Böhme, H. R. Kricheldorf, *Synthesis of LC Multiblock Copolymers* in: *ACS Symp. Series No.* **713** *Solvent-Free Polymerizations and Processes: Minimization of Conventional Organic Solvents*, M. Hunt, T. Long, Eds., American Chemical Society, Washington, DC (1998), pp. 8
10. D. Pospiech, L. Häußler, K. Eckstein, H. Komber, D. Voigt, A. Janke, A. Gottwald, D. Jehnichen, H. R. Kricheldorf, *LCP-Polysulfone Multiblock Copolymers: Combination of High Performance Polymers*, in:*Euromat'99, Proceedings* Vol. **13** „Functional Polymers", Munich (2000)
11. D. Pospiech, L. Häußler, D. Voigt, D. Jehnichen, A. Janke, A. Baier, K. Eckstein, F. Böhme, *J. Appl. Polym. Sci.* **62**, 11, 1819 (1996)
12. D. Pospiech, K. Eckstein, L. Häußler, H. Komber, D. Jehnichen, K. Grundke, F. Simon, *Macromol. Chem. Phys.* **200**, *6*, 1311 (1999)
13. P. Friedel, C. Kunert, D. Pospiech, A. Gottwald, D. Jehnichen, Application of mean field theory on the phase separation behavior of polydisperse Poly(sulfone-b-tetramethylene oxide) multiblock copolymers, submitted to *J. Polym. Sci., Polym. Phys. Ed.* (Dec., 1999)
14. L. Häußler, D. Pospiech, K. Eckstein, D. Jehnichen, *Thermochim. Acta* **339**, 87 (1999)
15. D. Pospiech, D. Jehnichen, L. Häußler, D. Voigt, K. Grundke, C. K. Ober, H. Körner, J. Wang, *Polymer Prepr., Am. Chem. Soc., Div. Polym. Chem.,* **39**, 2, 882 (1998)
16. D. Li, A. W- Neumann, *J. Colloid Interface Sci.* **148**, 190 (1992)
17. K. Grundke, D. Pospiech, W. Kollig, F. Simon, A. Janke, *Wetting on surfaces of heterogeneous block copolymers with fluorinated segments,* submitted to *Coll. & Polym. Sci.* (Apr., 2000)
18. J. A. de Fejter, *Biopolymers* **17**, 1759 (1978)

*Macromol. Symp.* ***2001**, 163, 127-134*



# Liquid Crystalline Triblock Copolymers. Mechanical Behaviour and Orientation under Uniaxial Strain

W. Gronski*, J. Sänger

Institut für Makromolekulare Chemie, Universität Freiburg, Stefan-Meier-Str. 31, D-79104 Freiburg, Germany

SUMMARY: The stress-strain and orientation behaviour of side-chain liquid crystalline(SLCP) ABA triblock copolymers with a backbone of polystyrene-*block*-1,2-polybutadiene-*block*-polystyrene and a cyanobiphenyl mesogen in the side chain was investigated in dependence of molecular weight. The polymer shows the behaviour of a thermoplastic liquid crystalline elastomer(TPLCE) in the nematic phase in a region between the glass transitions of the polystyrene block and the SLCP. The ultimate properties and E-modulus is lower than for conventional thermoplastic elastomers. Under uniaxial strain liquid crystalline order perpendicular to the direction of strain is induced.

## Introduction

ABA LC triblock copolymers with a low content of amorphous hard A-blocks and a soft LC center block represent a novel type of block copolymers that can be regarded as thermoplastic liquid crystalline elastomers (TPLCE). A TPLCE represents an anisotropic network and the question arises how the elastic properties of a TPLCE compare with those of conventional TPE's as well as with those of liquid crystalline elastomers (LCE). The synthesis and the morphological behaviour of TPLCE's based on polystyrene-b-1,2-polybutadiene-b-polystyrene triblocks has been reported in previous papers[1-2]. It has also been reported that the the TPLCE transforms to an optically transparent single-crystal-like material under large oscillatory shear[3]. In this paper the mechanical and orientational behaviour under uniaxial strain is reported.

## Materials

The materials studied in the present work are ABA triblock copolymers with polystyrene A-blocks and a side chain liquid crystalline polymer(SLCP) B-block of the following structure

 CCC 1022-1360/00/$ 17.50+.50/0

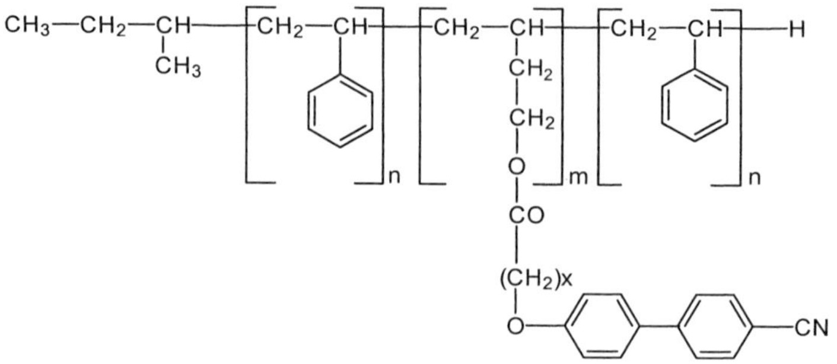

where x = 4 in the present case. Synthesis and characterization are have been described in a previous paper[1]. The phase transitions of the SLCP block are g/34°C/n/122°C/i. Triblocks of molecular weights 411000 g/mol (sample A) and 84000 g/mol (sample B) are investigated having a PS content of 10.5 wt% and 11.9 wt%, respectively. The block copolymers possess spherical bcc morphology in the isotropic melt[2] and the transition to the nematic phase induces a morphological transition into a hcp cylindrical morphology[2].

## Experimental

Block copolymers were melt pressed at 160°C for 20 min and cooled slowly to room temperature during 12h. After cooling the samples were annealed at 160°c for at least 24h. For stress-strain measurements standard samples according to DIN 53 455, form 4, were prepared. The measurements were carried out with a computer-controlled INSTRON 4204 testing machine at a stretching rate of 2 mm/s. Dynamic mechanical analysis (DMA) was carried out with a Rheometrix solid analyser RSAII in shear sandwhich geometry. 2D WAXS measurements were performed with a Kiessig camera and an image-plate detector system.

## Mechanical properties

In Fig. 1 stress-strain curves up to strain at break are shown for two molecular weights of the triblock copolymers. The materials exhibit rubber-like elasticity in a certain temperature range between Tg of the nematic center block and Tg of the PS domains. The typical plateau region of isotropic thermoplastic elastomers (TPE) as SIS block copolymers[4] is less developed and maximal strain and stress values are lower in the LC triblock copolymers. For the lower molecular weight material the tear strength decreases with increasing temperature until flow starts at about 64° C. In contrast the ultimate properties of the higher molecular weight material are almost independent of temperature up to 80°C and also the form of the

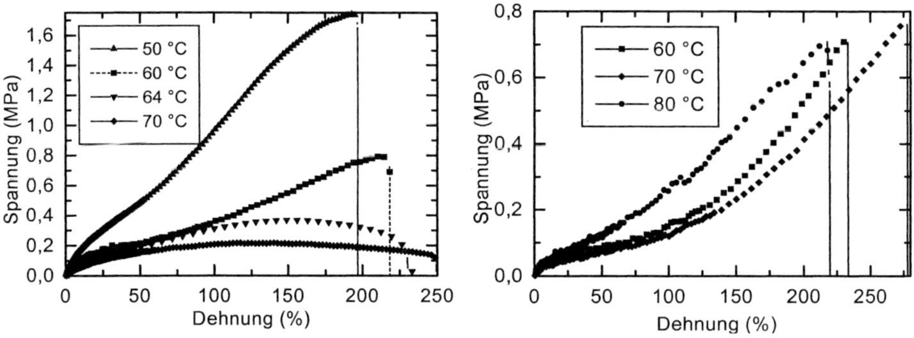


Figure 1: Stress-strain curves of triblock copolymer with Mn = 84 000 g/mol (left) and with Mn = 411 000 g/mol (right)

curves does not change and shows TPE-like behaviour. Within the limitation of the temperature range the higher molecular weight material may be considered as thermoplastic liquid crystalline elastomers (TPLCE). Cyclic stress-strain measurements shown in Fig. 2 show the typical hysteretic behaviour of TPE's. The permannt set is much more pronounced in the triblock copolymer of lower molecular weight. This is explained by the deformation of the PS cylindrical domains in addition to slow reorientation of the cylinders into the direction of strain. Permant set is diminished when the samples are annealed for some time at the measuring temperature. The initial state is completely restored after annealing at 160°C.

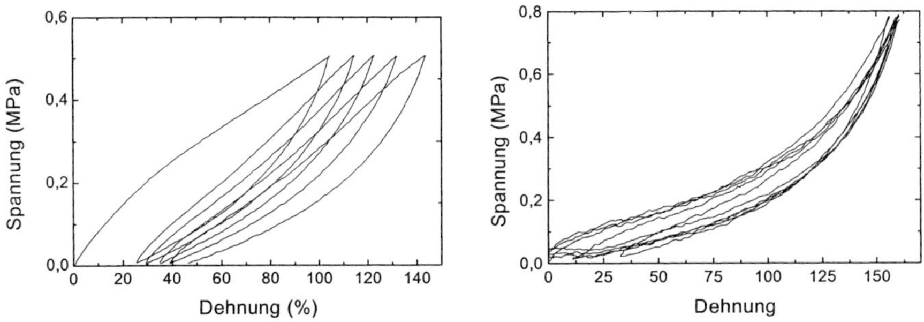


Figure 2: cyclic stress-strain of triblock with Mn = 84 000 g/mol (left) and with Mn = 411 000 g/mol (right)

Fig. 3 shows DMA curves measured with the higher molecular weight triblock copolymer. The pronounced drop of the shear modulus G' at the nematic-isotropic transition has also been observed for the homopolymer proving that the effect is not due to the morphological transition from cylindrical to spherical morphology that is induced by the phase transition. The effect is remarkable because in other nematic polymers only weak softening or even hardening was observed at $T_{ni}$[5,6]. The effect is reversible on heating and cooling.

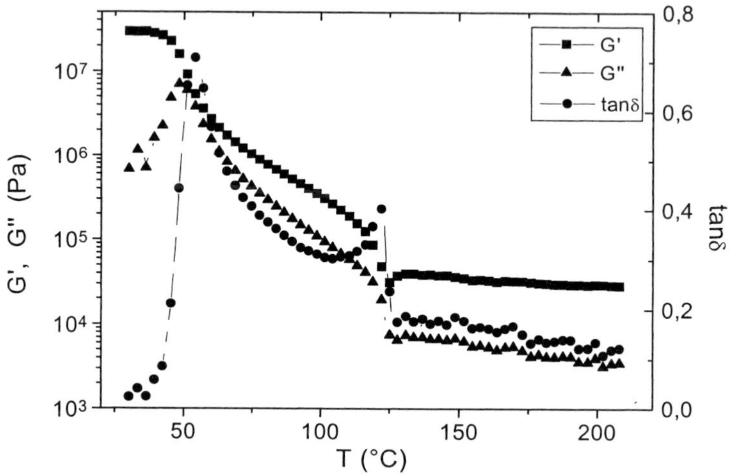


Figure 3: Dynamic mechanical behaviour of triblock with Mn = 411 000 g/mol

## Orientation under strain

In previous work it was shown that by large oscillatory shear that optically uniaxial transparent films were obtained having the properties of single liquid crystals[3]. It was shown that the nematic director and PS cylinders were oriented perpendicular to the shear plane having order parameters of 0.66 and 0.50, respectively. In order to determine the orientation behaviour under uniaxial strain the 2D WAXS pattern during stretching of a rectangular film at a temperature of 55°C was measured. Before data collection the sample was annealed for 30 min at the measurement temperature. Fig. 4 shows the intensity distribution of a sample strained to maximum strain.

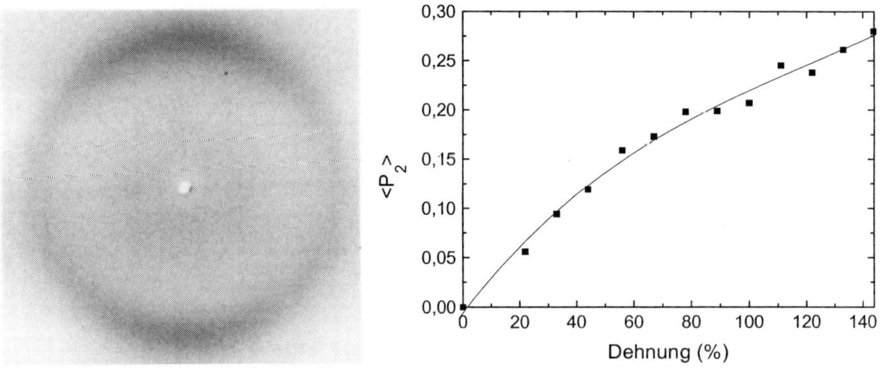


Figure 4: WAXS pattern at maximum extension(strain axis vertical) and order parameter as function of strain

The pattern shows that the mesogens are preferentially oriented perpendicular to the axis of strain and parallel to the film surface. A similar behaviour was found for chemically crosslinked nematic elastomers[7]. The orientation parameter $< P_2 >$ calculated from the azimuthal intensity distribution as a function of strain is also displayed. Order is reduced as compared to the order achieved under shear deformation. Also the sample remained turbid under uniaxial deformation, i. e. optical uniaxiality could not be achieved under these conditions as in the case of shear. At the present state we have no information about the orientation of the PS cylinders.

## Discussion

The preferred perpendicular orientation of the mesogens to the direction of strain indicates that the SLCP belongs to the class of LC polymers in which the main chain adopts an oblate chain configuaration with respect to the nematic director[8]. The additional orientation parallel to the film surface is explained by preferred homogeneous orientation of the nematogens at the film surfaces under condition of uniaxial strain(to be distinguished from planar shear where homeotropic anchoring in conjunction with the formation of smectic clusters is observed). As in the triblock copolymers a single-crystal-like state could not be achieved by uniaxial strain. Apparently the stretching of main chains along the strain axis on the one side and the perpendicular orienation of the director on the other side generates an unfavourabele entropic situation rich of defects.

Another reason of low LC order in the TPLCE's is the interaction of the LC phase with the PS cylinders. From the experiments with sheared block copolymers it is known that the long axes of the cylinders tend to be aligned along the nematic director, in other words homogeneous anchoring of the mesogens at the internal interfaces of the microdomains is preferred. When we assume that under uniaxil strain the cylinders are rotated into the direction of strain as it is the case for isotropic block copolymers with cylindrical morphology we see that there is a conflict between a tendency for parallelization of cylinders along the strain axis and the ordering tendency of the mesogens perpendicular to that direction. The results are not in contradiction to the behaviour under shear where both mesogens and cylinders were oriented perpendicular to shear direction and shear planes. The different behaviour is caused by a different orientation mechanism and different symmetry. Under shear the oblate chain conformations present in the LC polydomains are not only stretched but also rotated. If we suppose the rotation is the primary process and leads to preferential parallelization of the oblate chain configuation to the shear planes the interaction of the mesogens with the main chain will lead to perpendicular orientation of the director to the shear plane and also of the cylinders by minimization of the elastic energy.

In comparison to isotropic TPE's maximum elongation, E-modulus and tear strength are lower for the LC triblock copolymers. There are mainly two reasons that limit the extensibility of the materials. First one has to consider the anisotropy of the chain statistics which reduces chain extension. Second, and more important, the relation between the chain length to the molecular weight of the elastomeric block and to the critical chain length between entanglements is much smaller for SLCP's. Molar masses of mesogenic repeat units are about 6 – 9 times larger than molar masses of non-mesogenic units. Chain lengths are correspondingly shorter at a given molecular weight. Entanglement spacings of conventional elastomers correspond to degrees of polymerisation of the order of 100[9]. For typical SLCP's entanglement molecular weights of about $3x10^5$ g/mol were measured in the isotropic state corresponding to a degree of polymerization of about 730[6]. In the nematic state values between $8x10^5$ g/mol and $2.1x10^6$ g/mol were measured by different methods yielding DP's between 1940 and 5090. The DP values of the triblock copolymers of this work are only 200 and 1050, respectively. Since the molecular weight determining elastic properties of TPE's are mainly influenced by the entanglement molecular weight[4], the corresponding properties must be inferior in the LC triblock copolymers.

The comparatively low tear strength and E-modulus of the TPLCE's is connected to the properties of the PS-domains. The styrene content of the LC triblock copolymers is lower than in conventional TPE's which explains the reduced E-modulus. In conventional SIS triblock TPE's the tear strength is independent of the styrene content if the PS molecular weight is greater than 10 000 g/mol[4]. Although the molecular weight of the PS blocks of sample B is significantly lower than the threshhold value below which the tear strength starts to decrease for conventional TPE's, its tear strength is not lower than that of sample A the PS blocks of which surpasses that threshhold value two times. Therefore one concludes that the threshhold value is lower for the LC triblock copolymers due to a stronger tendency for microphase separation. The origin for lower tear strength of these materials must rather be seen in the closer distance to the glass transition of the PS-domains. Due to the molecular weight dependence of Tg flow sets in at lower temperature for the LC triblock copolymer having PS blocks of lower molecular weight.

The softening of the material at the nematic-isotropic transition is unusual with regard to results reported in literature. For nematic networks[5] a very small decrease or even an increase of the modulus was found[6]. The latter occurred for polymers with molecular weights above the entanglement molecular weight.$M_e$. No modulus change was found for molecular weight below $M_e$. The different behaviour observed for block copolymers and homopolymers in our case is thought to be due to the different response of the nematic director to shear. In the work of Rubin et al.[6] the mesogens were aligned parallel to the shear plane whereas in our case the director has a perpendicular orientation at large shear amplitudes. Even though only small shear amplitudes were applied in the DMA experiment and complete alignment is not expected we think that the greater modulus and viscosity in shear deformation is to be attributed to the tendency of the alignment of the mesogens paralallel to the gradient direction as in the previous experiments.

## Conclusion

It has been shown that under uniaxial strain the LC triblock copolymers have elastic properties in the nematic phase in a temperature range between the Tg of the amorphous PS blocks and the liquid crystalline center block. With the limitation of the temperature range of rubber-like behaviour they have properties of thermoplastic liquid crystalline elastomers(TPLCE). Ultimate properties and E-modulus are smaller than for conventional TPE's mainly due to the lower entanglement molecular weight of the SLCP. The orientation induced by strain is smaller than in chemically crosslinked elastomers because of an antagonistic orientation behaviour of the orientation of the mesogens transversal to the strain axis and of the orientation of the polystyrene cylindrical microdomains that orient parallel to the strain axis. The latter behaviour is an intrinsic property of the SLCP of the present system and may be different for a system having a prolate conformation. of the main chain.

## References


1. J. Sänger, W. Gronski, *Macromol. Chem. Phys.* **199**, 555 (1998)
2. J. Sänger, W. Gronski, S. Maas, B. Stühn, B. Heck, *Macromolecules* **30**, 6783 (1997)
3. J. Sänger, W. Gronski, H. Leist, U. Wiesner, *Macromolecules* **30**, 7621 (1997)
4. M. Morton in: *Thermoplastic Elastomers*, N.R. Legge, G. Holden Eds., Hanser Verlag, München 1987, p. 67 ff
5. W. Gleim, H. Finkelmann in: *Side chain liquid crystal polymers*, Blackie&Son, Glasgow, p. 287 (1987)
6. S. F. Rubin, R. M. Kannan, J. A. Kornfield, C. Boeffel, *Macromolecules* **28**, 3521 (1995)
7. J. Schätzle, H. Finkelmann, *Mol. Cryst. Liqu. Cryst.* **142**, 85 (1987)
8. M. Warner, K. P. Gelling, T. A. Vilgis, *J. Chem. Phys.* **88**, 4008 (1988)
9. W. W. Graessley, *Adv. Polym. Sci.* **16**, 1 (1974)

*Macromol. Symp.* ***2001,*** *163, 135-143*



# Bimodal Polyethylene —
# Interplay of Catalyst and Process

Frank P. Alt*, Ludwig L. Böhm, Hans-Frieder Enderle, Joachim Berthold

Elenac GmbH, R & D Frankfurt, Industriepark Höchst, 65926 Frankfurt am Main, Germany

SUMMARY: HDPE specialities having tailored bimodal molecular mass distributions suitable for applications as e.g. pipe or film material are formed in the Hostalen®-BM low pressure slurry process. A cascade of two STHD (stirred tank heavy diluent) reactors is used to produce at extremely different reaction conditions in each step of the polymerization an in situ polyethylene blend consisting of a low molecular mass homopolymer and a high molecular mass copolymer. This way, a resin having a very broad molecular mass distribution and an inverse comonomer distribution is obtained. High impact strength, stress crack resistance, stiffness, tensile strength, and good processability are combined and materials showing outstanding properties are achieved.

## Introduction

The properties of polyolefins depend mainly on the polymer structure. For instance, the rather simple monomer ethylene can be converted into a great variety of polymer grades showing completely different mechanical properties suitable for many applications.

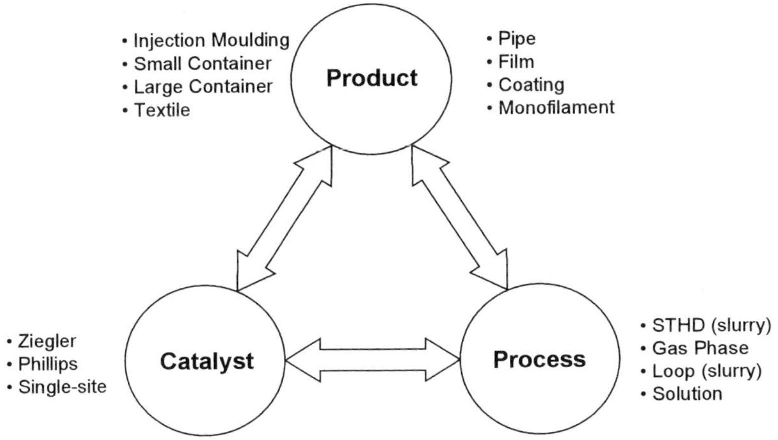


Fig. 1: Catalysts and processes for the production of polyethylenes.

 CCC 1022-1360/00/$ 17.50+.50/0

Much is known about the structure-property-relationship of polyethylene and numerous polymerization technologies have been developed in order to tailor the polymer architecture of polyethylene[1] (Fig. 1). In such a technology both parts, the process and the catalyst, have to form a well-balanced system[2]. The catalyst is crucial for the polymerization reaction. By the cooperation of process and catalyst a definite polymer structure is produced.

## Tailor-made bimodal molecular mass distribution

On principle, there are two ways to form polyethylene showing a bimodal molecular mass distribution (BMMD) (Fig. 2).

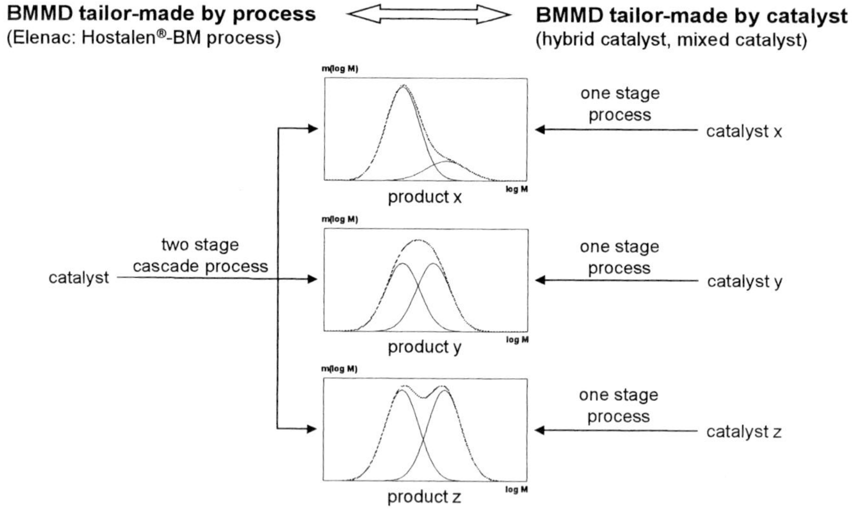


Fig. 2: Two ways towards bimodal polyethylenes.

One way is to synthesize a tailor-made catalyst. Hybrid catalysts or mixed catalysts unifying the properties of different active sites on each catalyst particle[3] can be used to obtain an in situ polyethylene reactor blend in a one stage process. For each bimodal grade a sophisticated catalyst has to be developed in order to adjust the amount, the molecular mass distribution, the comonomer incorporation, and the comonomer distribution of both fractions formed in the polymerization reaction.

Another way towards bimodal polyethylene in situ blends is to use a two stage cascade polymerization process. The catalyst passes two reaction zones and one polymer fraction is formed after the other at different reaction conditions. Only one optimized catalyst is required for the production of various grades. The desired molecular mass distribution and comonomer distribution are designed by the process. Beside these features this technology offers further advantages: No change of the catalyst or cocatalyst is necessary when the grade is switched. Change is done by adjusting the reaction conditions whilst the space-time-yield remains constant. Thus, little time is required to pass on to the next grade and the amount of offspec material produced during the switch is nearly negligible. Starting from the catalyst feed and ending with the pelletized final product the cascade process can be designed as one continuously working production line.

## STHD cascade process

The concept of a two stage cascade process is for instance realized in Elenac's Hostalen®-BM low pressure slurry technology. The core of this continuous polymerization process consists of two stirred tanks in series. Only one highly sophisticated 3rd generation Ziegler catalyst is used to produce all bimodal grades (Fig. 3).

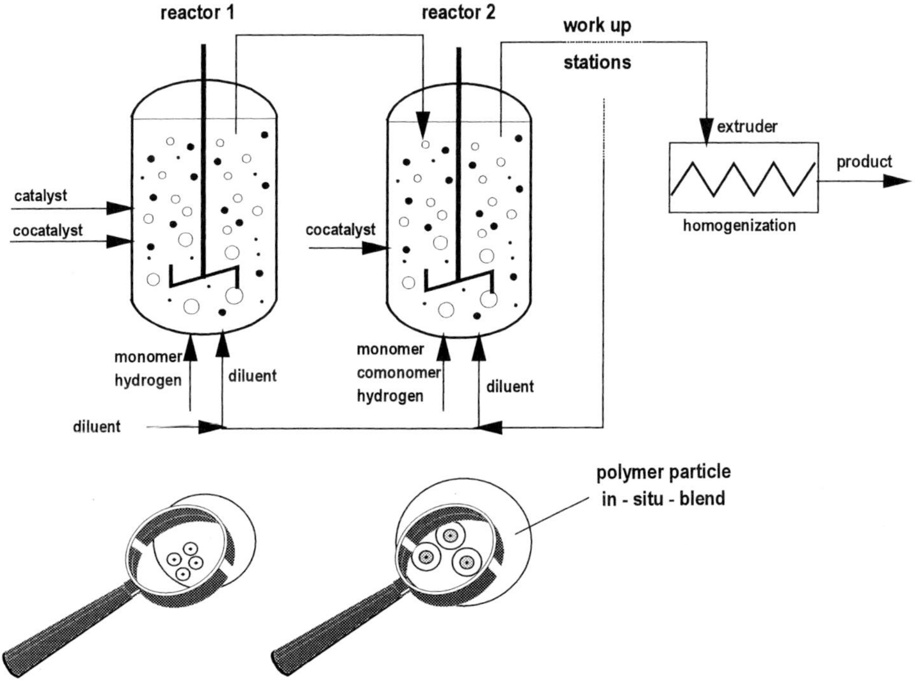


Fig. 3: Elenac's Hostalen®-BM low pressure slurry process.

In the first reactor a very low molecular mass ethylene homopolymer is formed. The catalyst, cocatalyst, ethylene, and diluent (e.g. hexane) are continuously fed to reactor 1. The molecular mass of this polymer fraction is adjusted by adding an appropriate amount of hydrogen. Then the slurry is transferred into reactor 2 where the polymer particles with the still active catalyst polymerize under completely different reaction conditions and comonomer is added. A very high molecular mass copolymer is produced in this second stage. The polymer slurry is then passed to the working up station. The polymer powder is separated, dried, compounded and pelletized. The diluent is recycled after purification.

Every particle of the polymer powder produced in the cascade process has the same average composition. In the particle forming process[4, 5] each catalyst particle has to have the same activity and every volume increment has to contain the same number of active sites. Thus, the growth of the particles during the polymerization results in polymer grains showing evenly distributed layers of low molecular mass homopolymer and high molecular mass copolymer around the catalyst fragments[1] (Fig. 3 and Fig. 4).

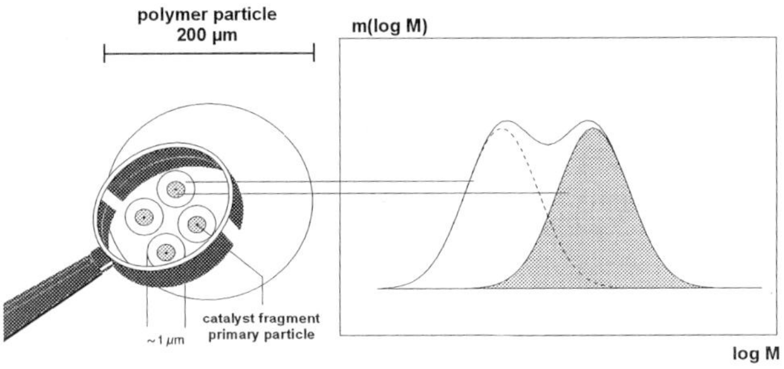


Fig. 4: Polymer particle structure of an in situ polyethylene blend produced by a cascade process.

This way, an in situ polyethylene blend is obtained. Such a high performance polymer alloy shows an outstanding combination of desired properties for e.g. pipe and film applications. The mechanical data measured (e.g. stress crack resistance, impact strength) are superior to those that can be achieved with conventional blends.

# Structure-property-relationship of bimodal polyethylene

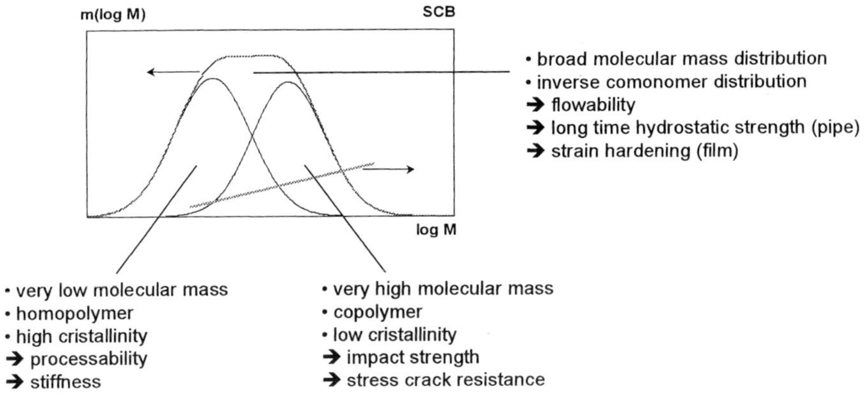


Fig. 5: Structure-property-relationship of bimodal polyethylene.

By the synthesis of bimodal polyethylenes the mechanical properties of the low molecular mass fraction and the high molecular mass fraction are combined (Fig. 5). This allows to produce materials showing higher stiffness, higher toughness, improved impact strength, higher stress crack resistance, and optimized processability compared to unimodal resins. The inverse comonomer distribution (Fig. 6 and Fig. 7) results in products with longer lifetime under stress[6, 7, 8, 9].

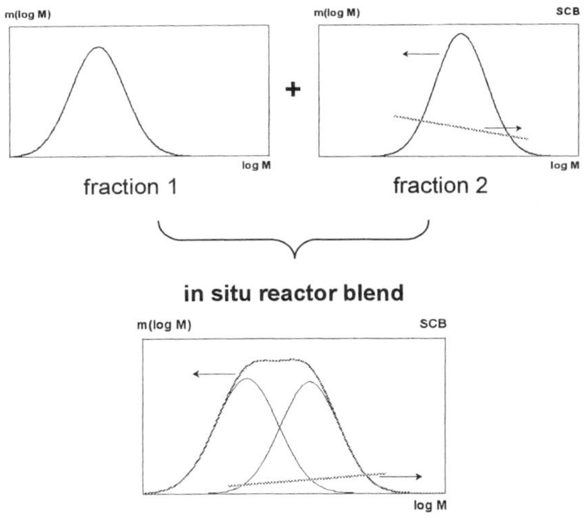


Fig. 6: Inverse comonomer distribution of an in situ reactor blend.

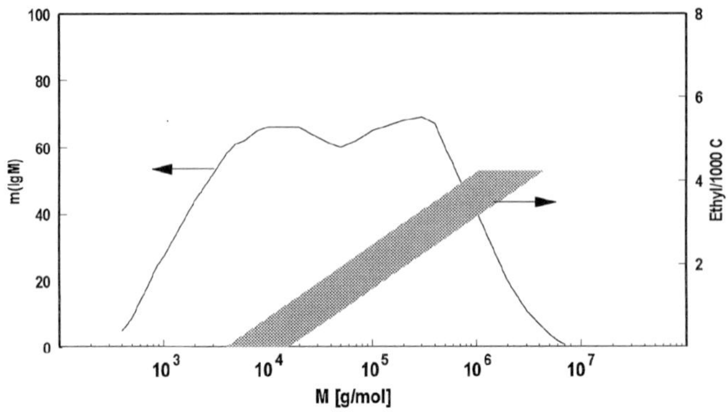


Fig. 7: Molecular mass distribution (SEC) and comonomer distribution (Holtrup fractionation, FT-IR analysis) of a bimodal PE 100 pipe material.

Due to the fact that the comonomer is mainly incorporated into the long polymer chains bimodal polyethylene has a higher density compared to an unimodal polyethylene showing the same amount of short chain branches but no inverse comonomer distribution (Fig. 8). For that reason, bimodal polyethylene grades exhibit higher stiffness and enhanced stress crack resistance[10)].

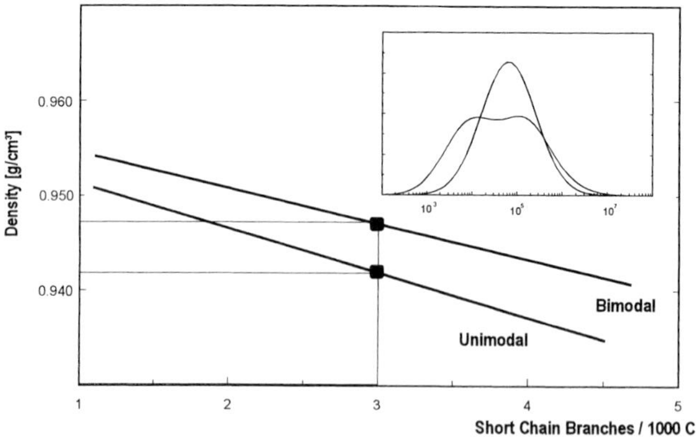


Fig. 8: Density of unimodal and bimodal polyethylene as a function of short chain branches.

# Catalysts optimized for the STHD cascade process

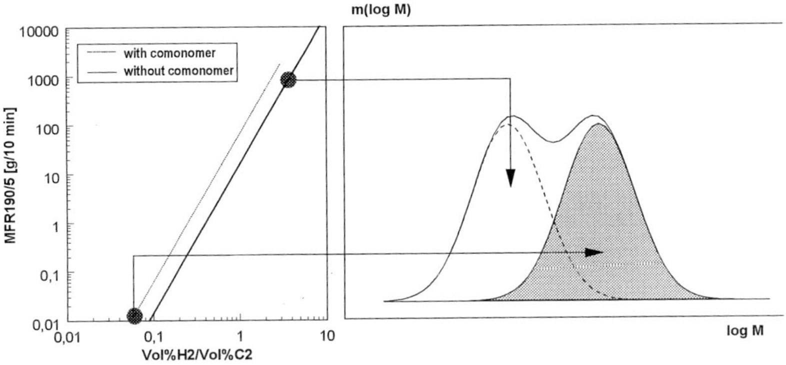


Fig. 9: Molecular mass as a function of hydrogen and comonomer concentration in a cascade process using an unimodal catalyst tailor-made for the production of bimodal polyethylene.

When the growing catalyst particles in the slurry pass through the reactor cascade they are subjected to extremely different reaction conditions. The molecular mass of the two fractions is controlled by the concentration of hydrogen within the vessel. Therefore, the catalyst has to show an excellent hydrogen response (Fig. 9) to reduce the amount of hydrogen required for molecular mass regulation because hydrogen decreases the activity of the catalyst[1].

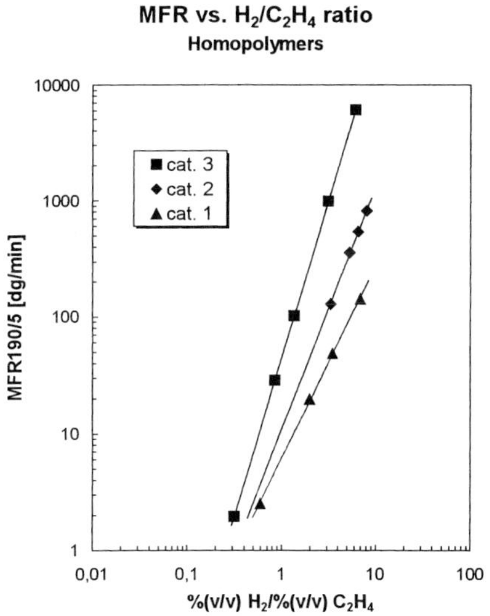


Fig. 10: Optimization of the catalyst with regard to hydrogen response.

The catalyst has further to be optimized with regard to the broadness of the molecular mass distribution and the comonomer distribution of the polymer produced. Also, the copolymerization behavior (Fig. 9) and the comonomer incorporation rate (Fig. 6) has to be to taken into consideration. As well, the catalyst has to show a very high productivity and activity in the polymerization reaction. Finally, to tailor products by the process a detailed knowledge about the temperature dependence of all these parameters is required.

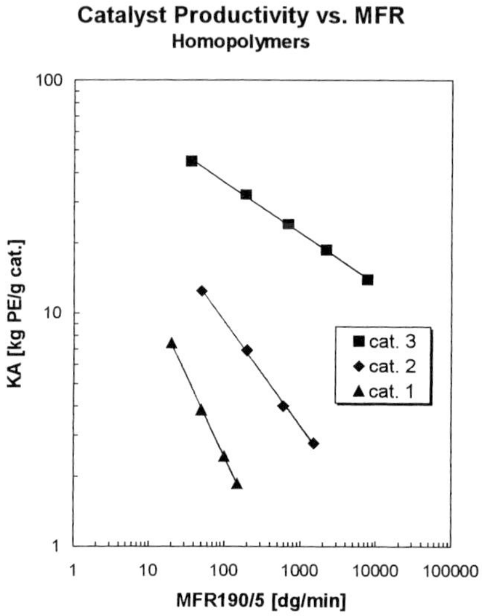


Fig. 11: Catalyst productivity as a function of melt flow rate.

In the particle forming process every polymer grain originates from one catalyst particle. Provided that no fragmentation of the catalyst particles into smaller sub-particles occurs during the polymerization and that every catalyst particle has the same productivity and activity the catalyst and the polymer powder show the same broadness of the particle size distribution[4, 5)].

Ziegler catalysts are most suitable for Elenac's Hostalen®-BM low pressure slurry process. Highly sophisticated Ziegler catalysts of the 3rd generation supported on $MgCl_2$ have been developed. Fig. 10 and Fig. 11 show the improvement of the catalyst with respect to the hydrogen response and the productivity. Polymer powders having a high bulk density and a narrow particle size distribution are obtained. The catalyst shows an excellent hydrogen response in combination with high activity and productivity and is capable of forming a homogeneous comonomer distribution.

By the combination of this optimized catalyst and the cascade slurry process products with bimodal molecular mass distribution exhibiting outstanding properties are obtained.

## References


1. L. L. Böhm, J. Berthold, H.-F. Enderle, M. Fleissner, in: *Metalorganic Catalysts for Synthesis and Polymerization*, W. Kaminsky (Ed.), Springer-Verlag, Berlin 1999, p. 3ff
2. T. Xie, K. B. McAuley, J. C. C. Hsu, D. W. Bacon, *Ind. Eng. Chem. Res.* **33**, 449 (1994)
3. Th. J. Pullukat, R. E. Hoff, *Catal. Rev. – Sci. Eng.* **41(3&4)**, 389 (1999)
4. L. L. Böhm, *Chem. Ing. Techn.* **56**, 674 (1984)
5. L. L. Böhm, *DECHEMA Monograph.* **131**, 191 (1995)
6. J. Berthold, L. L. Böhm, H.-F. Enderle, P. Göbel, H. Lüker, R. Lecht, U. Schulte, *Plastics, Rubber and Composites Proc. & Appl.* **25**, 368 (1996)
7. L. L. Böhm, H.-F. Enderle, M. Fleissner, in: *Catalyst Design for Tailor-Made Polyolefins*, K. Soga, M. Terano (Eds.), Kodansha Tokyo 1994
8. U. Schulte, R. Lecht, *Swiss Plastics* **20**, 15 (1998)
9. U. Schulte, *Kunststoffe* **87**, 203 (1997)
10. J. Schreirs, L. L. Böhm, J. C. Boot, P. S. Leevers, *TRIP* **4**, 408 (1996)

*Macromol. Symp.* ***2001****, 163, 145-156*


# Controlled Radical Polymerization in the Presence of N-Oxyls

Susanne Brinkmann-Rengel,* Heinz-Friedrich Sutoris, Horst Weiß

BASF AG, 67056 Ludwigshafen, Germany

SUMMARY: Recent development in controlled radical polymerization has provided a tool to combine a relatively robust radical polymerization technique with structural control. This contribution focuses on stable free radical polymerization in the presence of nitroxides. The influence of 2,2,6,6-tetra-methyl-piperidine-N-oxyl (TEMPO) and temperature on the copolymerization of styrene and acrylonitrile will be discussed. In the second part a new class of nitroxide stable free radicals will be presented that shows enhanced performance in styrene polymerizations.

## Introduction

Industrial plastics can be divided into commodity, engineering and high performance plastics. Unlike the prediction made in 1975 performance plastics took less than 1% of the total market share in 1998. This trend indicates that the development of new polymers from new monomers is not crucial. Further development focuses mainly on property improvement of existing products. In this context the synthesis of defined materials from existing monomers has proven more and more important.

The design of defined materials includes molecular architecture, monomer selection, and introduction of functionality. Polymer molecules can be designed to linear, comb, star-shaped, dendritic, highly branched or network structures. If only one monomer is used the resulting polymers can be distinguished by molecular weight, polydispersity, or tacticity. If more than one monomer is applied versatility is further increased to statistical, alternating, tapered, block or graft copolymers. Last but not least functional groups can be incorporated into the polymer chain. Functional groups can be located at one or both chain ends or along the chain. Thus a variety of materials can be designed from existing components - provided a tool to control synthesis is available.


CCC 1022-1360/00/$ 17.50+.50/0

Radical polymerization is an important tool in polymer synthesis, because it tolerates a large number of functional groups and the polymerization process can easily be handled in large scales. Recently developed methods to control the polymerization are valuable as the versatility of radical polymerization can be combined with structural control. Several approaches to controlled radical polymerization (CRP) are reported in literature. The most important are stable free radical polymerization (SFRP), atom transfer radical polymerization (ATRP) and reversible addition fragmentation chain transfer (RAFT).[1-5] The advantages of the different methods are outlined elsewhere.[6] This contribution focuses on SFRP mediated by N-oxyl compounds.

## Nitroxide mediated Stable Free-Radical Polymerization

The development of nitroxide-mediated polymerization was first performed by the groups of Solomon and Rizzardo and Georges.[7,8] A stable free radical is introduced into the polymerization medium and intervenes the polymerization process by establishing an equilibrium between growing polymer chains and stable free radical and polymer-N-oxyl adducts. The equilibrium is shifted towards the adducts which can be activated thermally. The concentration of growing polymer chains in the reaction medium is reduced. Biradical termination can be suppressed so that the growth of polymer chains can be well directed.

Scheme 1: Stable Free Radical Polymerization in the Presence of TEMPO.

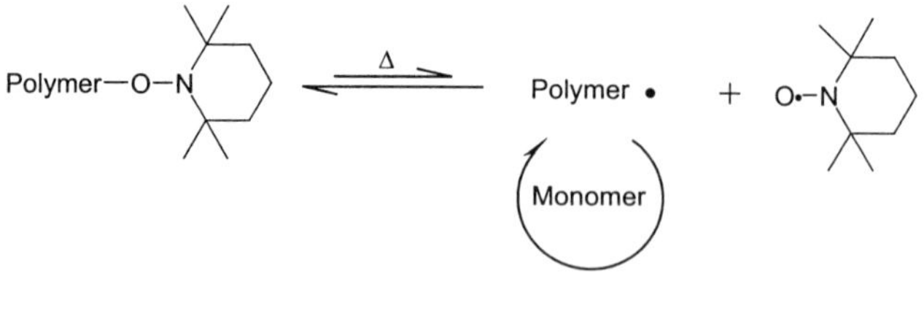


Originally 2,2,6,6-tetra-methyl-piperidine-N-oxyl (TEMPO) was used as stable free radical. Due to the nature of the polymer-TEMPO bond polymerization is limited to reaction temperatures above 100 °C. Due to the reduced concentration of growing chains polymerization rate is reduced, too. Although termination reactions can be diminished, termination occurs eventually and free TEMPO accumulates in the reaction medium. Due to this persistent radical effect the polymerization slows down and finally stops.[9] Thus full conversion cannot be obtained and accessible molecular weights are limited. Only styrene and its derivatives or mixtures can be polymerized successfully.[10, 11]

Research of several groups aimed to overcome the problems outlined above. Work focused on the one hand on the investigation of kinetics and on comprehending the mechanism. [10, 12-16] Additives were introduced to speed up polymerization. [17-23] In another approach new stable free radicals were developed to enable polymerization at lower temperatures using a variety of monomers.[24-31] Calculations can be used to estimate the stability of N-alkoxyamine bonds.[32]

In this contribution two examples of our work will be briefly outlined. In the first part the influence of TEMPO on the copolymerization of styrene and acrylonitrile will be discussed. The second part focuses on a new class of controller molecules that shows enhanced polmerization performance in comparison to TEMPO.

**Influence of TEMPO on Controlled Radical Copolymerization**

Until recently only a few works dealt with controlled radical copolymerization.[33-42] We chose styrene and acrylonitrile as model system. The copolymerization of styrene and acrylonitrile was first studied in 1996 by Fukuda et al.[42] As his work focused on the preparation of poly(styrene)-*block*-poly(styrene-co-acrylonitrile) block copolymers, the copolymerization itself was not investigated in detail. In the mean time additional work was reported in literature.[43, 44] Although TEMPO acts as an inhibitor in acrylonitrile homopolymerization, SAN copolymers can be prepared by CRP. As pointed out above, the TEMPO-polymer bond can be activated thermally. The temperatures necessary for controlled radical polymerization are significantly higher than many examples of conventional radical polymerization reported in literature. The objective of our work was to investigate the influence of TEMPO and temperature on the copolymerization of styrene and acrylonitrile.[45, 46]

When molecular weights and polydispersities are compared, the differences between materials prepared in the presence or absence of TEMPO are obvious.

Whereas molecular weights above 300.000 g/mol and polydispersities around 2.0 are obtained when azeotropic mixtures of SAN are polymerized following a conventional procedure (FRP) using either thermal or dibenzoylperoxide initiation, molecular weights and polydispersities of samples prepared in the presence of TEMPO are significantly lower. Polydispersities range from 1.2 to 1.4 and molecular weights grow linearly as a function of conversion until conversions of about 60 % are reached. At higher conversion molecular weights level off to smaller values and polydispersity increases due to inevitable termination reactions. An induction period is observed when TEMPO is used as mediator.

Table 1: TEMPO-mediated Polymerization of Styrene and Acrylonitrile (75 : 25).

| 110 °C | | | | 120 °C | | | |
|---|---|---|---|---|---|---|---|
| Time [h] | Conversion [%] | Mn (GPC) [g/mol] | PDI | Time [h] | Conversion [%] | Mn (GPC) [g/mol] | PDI |
| 1 | 2 | -- | -- | 1 | 1 | -- | -- |
| 2 | 3 | 7,300 | 1,26 | 2 | 4 | 8,631 | 1,33 |
| 5 | 33 | 33,000 | 1,38 | 4 | 39 | 36,580 | 1,36 |
| 8 | 50 | 44,600 | 1,39 | 8 | 65 | 51,210 | 1,43 |
| 16 | 65 | 55,100 | 1,40 | 16 | > 90 | 56,610 | 1,62 |
| 40 | 85 | 58,800 | 1,53 | - | - | - | - |

To investigate the influence of TEMPO on copolymerization the acrylonitrile content of the monomer feed was varied over the whole composition range. Polymerizations were stopped below 5 percent conversion to rule out composition drift during polymerization. No major difference in the styrene content of polymers prepared by conventional or controlled radical polymerization was found.

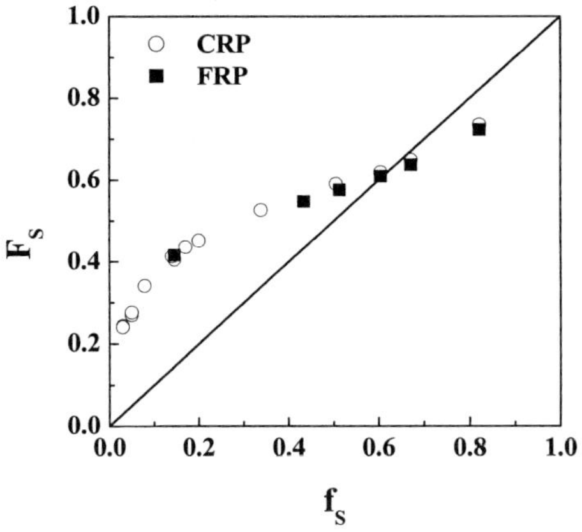


Figure 1: Copolymerization Diagram for Polymerizations carried out at 120 °C.

The copolymerization parameters were calculated using the method described by Kelen and Tüdös for temperatures ranging from 95 to 135 °C.[47] Taking the margin of errors into account no major difference between controlled and conventional radical polymerization is observed. The copolymerization parameters remain almost constant over the temperature range investigated. Nevertheless, it has to be noted that the temperature range for controlled radical copolymerization of styrene and acrylonitrile is limited: Below approx. 100 °C the polymerization is inhibited, above 135 °C control is lost using the controller concentration reported here.

Table 2: Apparent Copolymerization Parameters.

| | FRP | | CRP | |
|---|---|---|---|---|
| *T* *[°C]* | $r_{AN}$ | $r_S$ | $r_{AN}$ | $r_S$ |
| 97 ± 3 | 0.07 | 0.37 | 0.07 | 0.43 |
| 110 ± 3 | 0.07 | 0.41 | 0.08 | 0.45 |
| 122 ± 3 | 0.07 | 0.40 | 0.08 | 0.45 |
| 130 ± 3 | 0.07 | 0.41 | 0.09 | 0.53 |

Although temperature dependence of copolymerization parameters derived from the terminal model is not observed, Ferrando and Longo recently reported evidence for temperature dependence of copolymerization parameters derived from the penultimate model, which result in a drift in sequence distribution with temperature.[48] The sequence distribution of polymers prepared at 120 °C was investigated and compared to the results obtained from Ferrando and Longo and to predictions from terminal and penultimate model calculated for the temperature investigated. The results obtained in our lab compare well with Ferrando and Longo, no difference between FRP and CRP is observed.[46] We additionally performed Monte-Carlo and PREDICI simulations, which will be reported elsewhere.[49]

## New Nitroxides for Stable Free Radical Polymerization

The development of new controllers is an important tool to overcome limitations of the TEMPO-mediated process. To influence the TEMPO-polymer bond bulky substituents are introduced into the vicinity of the nitrogen atom. Aditionally heteroatoms are incorporated into the controller molecule, as different electron withdrawing or donating groups could influence the TEMPO-polymer bond as well. Three N-oxyls of the morpholine and morpholone type were synthesized and evaluated in styrene polymerization.

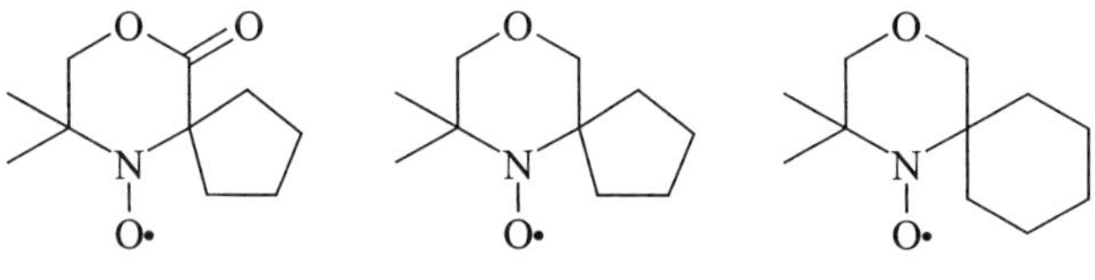


**1** **2** **3**

As shown in scheme 2, the synthesis of the controller molecules is accomplished following a route reported in literature for the synthesis of hindered amines.[50] The base induced reaction of aminopropanol, chloroform and a ketone yields a morpholone that can be directly oxidized to the N-oxyl. Alternatively, the morpholone is reduced, submitted to a ring-closing reaction followed by oxidation. The second route results in morpholine type N-oxyls.

Scheme 2: Synthesis of Morpholone- and Morpholine-N-Oxyls.

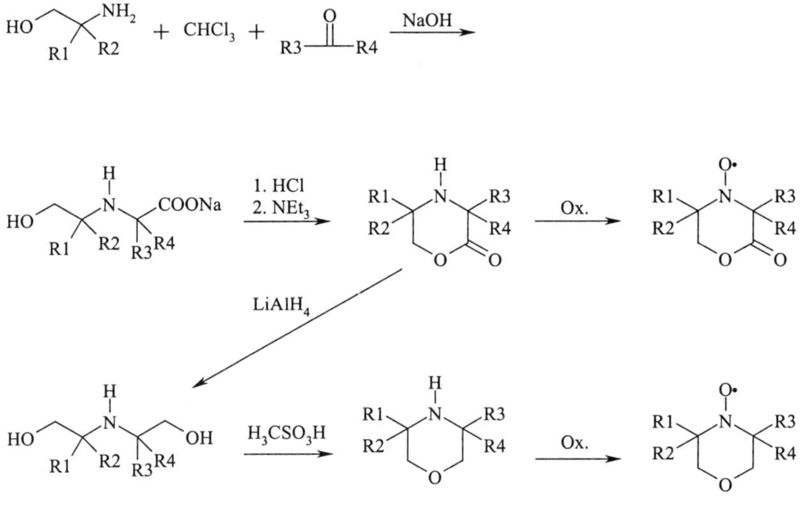


Nitroxides **1**, **2**, and **3** were used as mediators in styrene polymerizations performed at 125 °C in bulk. Using **1**, **2** and **3** higher conversions are obtained in shorter times. Using nitroxide **3** conversions larger 55 % can be obtained in two hours. Polymerizations performed in the presence of TEMPO yield 47 % conversion in five hours. No induction period is observed when **1**, **2**, or **3** is used as in the case of TEMPO. It is not immediately clear which structural feature in nitroxides **1**, **2**, or **3** reduces the induction period. Further investigations have to explore this aspect.

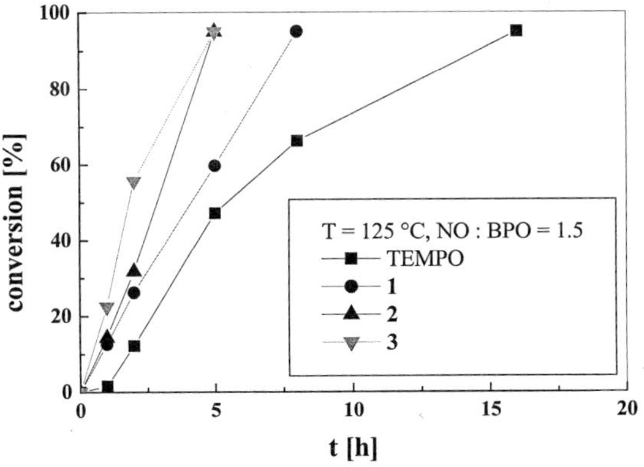


Figure 2: Evolution of Conversion as a Function of Reaction Time.

Polymers obtained in the presence of the new nitroxides show an almost linear dependence of molecular weight on conversion. Molecular weights observed compare well with results obtained from the TEMPO-mediated process.

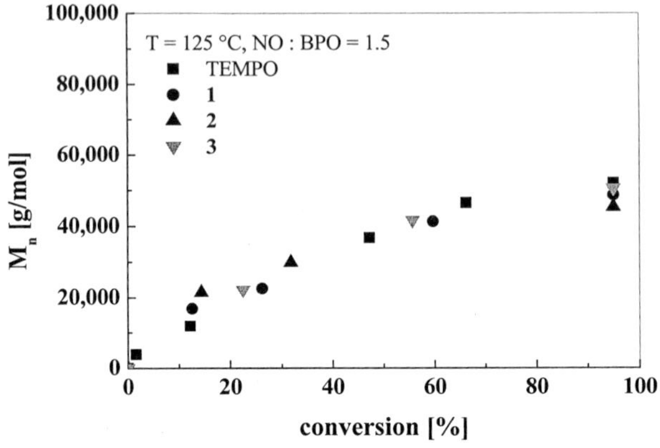


Figure 3: Molecular Weight as a Function of Conversion.

The evolution of polydispersity with conversion shows however, that accelerating polymerization results in larger polydispersities compared to TEMPO. Interestingly, polymers prepared in the presence of **3** that showed the fasted polymerization performed also best in terms of polydispersity.

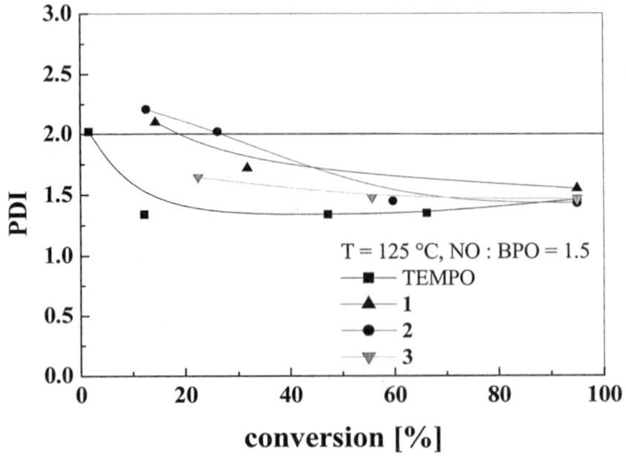


Figure 4: Polydispersity as a Function of Conversion.

The resulting polystyrene polymers can be used as macroinitiators in n-butyl acrylate polymerization. The polystyrene starting material prepared in the presence of **1** is dissolved in n-butyl acrylate and heated at 120 °C. After four hours 15 % conversion are observed. The number average molecular weight of the isolated polymer is shifted from 38,200 g/mol (PS-starting material) to 47,300 g/mol. The poly(n-butyl acrylate) content is estimated to be 22 % by $^{1}H$ NMR. Differential scanning calorimetry shows two glass transition temperatures at –46 and +98 °C, respectively. Recently Puts et al. outlined the advantages of nitroxides similar to **1** in acrylate polymerization.[51, 52] Improved performance from heteroatom containing nitroxides was recently published by Chong and co-workers.[29]

## Conclusion

A variety of materials can be obtained from existing starting materials when the polymerization process can be controlled to yield defined materials. Through the combination of a robust radical polymerization technique tolerating many functional groups with enhanced structural control during polymerization, stable free radical polymerization is a valuable tool for the synthesis of defined materials.

The TEMPO molecule does not influence the copolymerization behavior of styrene and acrylonitrile in the temperature range investigated. Composition, copolymerization parameters and sequence distribution are within the margin of error compared to conventionally prepared polymers. Nevertheless, when materials prepared by conventional and controlled radical copolymerization are compared, differences may occur, which do not only result from differences in molecular weight and polydispersity. When copolymers are prepared via conventional radical polymerization and the azeotropic monomer composition is not used (and the composition drift is not compensated for by monomer feed) polymer chains with different composition among the polymer chains are obtained. In controlled radical polymerization however, the majority of chains is initiated at the same time resulting in polymers with a composition drift along the chains.

Bulky substituents and heteroatoms in the vicinity influence the stable free radical-polymer bond and thus the performance and outcome of polymerizations. This displays an important tool to overcome limitations of the TEMPO system.

## Experimental

Monomers (styrene, acrylonitrile, n-butyl acrylonitrile), dibenzoylperoxide, and 2,2,6,6-tetra-methyl-piperidine-N-oxyl were used as supplied.

Polymerizations were performed in bulk using sealed glass ampoules protected by metal containers. The ratio of monomer to mediator to initiator was 1400 : 1.5 : 1 on a molar basis. Polymerizations were stopped by quenching to room temperature. The resulting polymers were precipitated into methanol, washed with methanol, and dried in vacuum.

Conversion was determined gravimetrically using a Mettler HR73 Moisture Analyzer. Molecular weights were determined by size exclusion chromatography using THF as eluent and polystyrene calibration. The acrylonitrile and butyl acrylate content was determined by NMR.

## Acknowledgements

Support from D. Christie, Y. Durant is gratefully acknowledged. J. Malchus and Y. Cao are acknowledged for performing simulations. We thank K. Knoll, S. Lehmann, I. Hennig, W. Heckmann for analytics and N. Nießner, S. Haremza, B. Raether, and P. Barghoorn for helpful discussions.

## References

1 M. K. Georges, R. P. N. Veregin, P. M. Kazmaier, G. K. Hamer *TRIP* **2**, 66-72 (1994)
2 C. J. Hawker *Acc. Chem. Res.* **30**, 373-382 (1997)
3 T. E. Patten, K. Matyjaszewski *Adv. Mat.* **10**, 901-915 (1998)
4 J. Chiefari, Y. K. Chong, F. Ercole, J. Krstina, J. Jeffery, T. P. T. Le, R. T. A. Mayadunne, G. F. Meijs, C. L. Moad, G. Moad, E. Rizzardo, S. Thang *Macromolecules* **31**, 5559-5562 (1998)
5 Y. K. Chong, T. P. T. Le, G. Moad, E. Rizzardo, S. H. Thang *Macromolecules* **32**, 2071-2074 (1999)
6 B. Raether *submitted to this journal*
7 US Patent 4,581,429, March 27,1985, D. H. Solomon, E. Rizzardo, P. Cacioli
8 M. K. Georges, R. P. N. Veregin, P. M. Kazmaier, G. K. Hamer *Macromolecules* **26**, 2987-2988 (1993)
9 H. Fischer *Macromolecules* **30**, 5666-5672 (1997)

10 D. Greszta, K. Matyjaszewski *Macromolecules* **29**, 5239-5240 (1996)
11 D. Greszta, K. Matyjaszewski *Macromolecules* **29**, 7661-7670 (1996)
12 J. M. Catala, F. Bubel, S. O. Hammouch *Macromolecules* **28**, 8441-8443 (1995)
13 R. P. N. Veregin, P. G. Odell, L. M. Michalak, M. K. Georges *Macromolecules* **29**, 3346-3352 (1996)
14 K. Ohno, Y. Tsujii, T. Fukuda, *Macromolecules* **30**, 2503-2506 (1997)
15 A. Goto, T. Fukuda *Macromolecules* **30**, 5183-5186 (1997)
16 J. He, H. Zhang, J. Chen, Y. Yang *Macromolecules* **30**, 8010-8018 (1997)
17 M. K. Georges, R. P. N. Veregin, P. M. Kazmaier, G. K. Hamer, M. Saban *Macromolecules* **27**, 7228-7229 (1994)
18 P. G. Odell, R. O. N. Veregin, L. M. Michalak, D. Brousmiche, M. K. Georges *Macromolecules* **28**, 8453-8455 (1995)
19 R. P. N. Veregin, P. G. Odell, L. M. Michalak, M. K. Georges *Macromolecules* **29**, 4161-4163 (1996)
20 M. V. Baldovi, N. Mohtat, J. C. Scaiano *Macromolecules* **29**, 5497-5499 (1996)
21 D. Greszta, K. Matyjaszewski *J. Polym. Sci. A: Polym. Chem.* **35**, 1857-1861 (1997)
22 B. Keoshkerian, M. K. Georges, M. Qinlan, R. Veregin, B. Goodbrand *Macromolecules* **31**, 7559-7561 (1998)
23 H. Baethge, S. Butz, C.-H. Han, G. Schmidt-Naake *Angew. Makromol. Chem.* **267**, 52-56 (1999)
24 D. Benoit, S. Grimaldi, J. P. Finet, P. Tordo, M. Fontanille, Y. Gnanou *Polymer Preprints* **38 (1)**, 729-730 (1997)
25 R. D. Puts, D. Y. Sogah *Macromolecules* **29**, 3323-3325 (1996)
26 S. Butz, H. Baethge, G. Schmidt-Naake *Angew. Macromol. Chem.* **270**, 42-48 (1999)
27 Y. Miura, S. Mibae, H. Moto, N. Nakamura, B. Yamada *Polymer Bulletin* **42**, 17-24 (1999)
28 D. Benoit, V. Chaplinski, R. Braslau, C. J. Hawker *J. Am. Chem. Soc.* **121**, 3904-3920 (1999)
29 Y. K. Chong, F. Ercole, G. Moad, E. Rizzardo, S. H. Thang *Macromolecules* **32**, 6895-6903 (1999)
30 D. Benoit, E. Harth, P. Fox, R. M. Waymout, C. J. Hawker *Macromolecules* **33**, 363-370 (2000)
31 S. Grimaldi, J.-P. Finet, F. Le Moigne, A. Zeghdaoui, P. Tordo, D. Benoit, M. Fontanille, Y. Gnanou *Macromolecules* **33**, 1141-1147 (2000)
32 P. Marsal, M. Roche, P. Tordo, P. de Sainte Claire *J. Phys. Chem.* **103**, 2899-2905 (1999)
33 C. J. Hawker, E. Elce, J. Dao, W. Volksen, T. P. Russell, G. G. Barclay *Macromolecules* **29**, 2686-2688 (1996)
34 G. Schmidt-Naake, S. Butz *Macromol. Rapid Commun.* **17**, 661-665 (1996)
35 N. Ide, T. Fukuda Macromolecules 30, 4268-4271 (1997)
36 H. Baethge, S. Butz, G. Schmidt-Naake *Macromol. Rapid Commun.* **18**, 911-916 (1997)
37 S. Butz, H, Baethge, G. Schmidt-Naake *Macromol. Rapid Commun.* **18**, 1049-1055 (1997)

38 N. Ide, T. Fukuda *Macromolecules* **32**, 95-99 (1999)
39 H. Baethge, S. Butz, C. Han, G. Schmidt-Naake *Angew. Makromol. Chem.* **267**, 52-56 (1999)
40 E. Yoschida, Y. Takiguchi *Polymer J.* **31**, 429-434 (1999)
41 C. H. Han, M. Drache, G. Schmit-Naake *Angew. Makromol. Chem.* **264**, 73-81 (1999)
42 T. Fukuda, T. Terauchi, A. Goto, Y. Tsujii, T. Miyamoto *Macromolecules* **29**, 3050-3052 (1996)
43 M. Baumert, Diploma Thesis, Albert-Ludwig-Universität, Freiburg, Germany, 1996
44 M. Baumann, Diploma Thesis, Technische Universität Clausthal, Clausthal-Zellerfeld, Germany, 1997
45 S. Brinkmann-Rengel, J. Cao, N. Nießner *Polymer Preprints* **40 (2)**, 362-363 (1999)
46 S. Brinkmann-Rengel, N. Nießner *submitted to ACS Symp. Ser.*
47 T. Kelen, F. Tüdös *J. Macromol. Sci. – Chem.* **A9**, 1-27 (1975)
48 A. Ferrando, A. Longo *Polymer Preprints* **38 (1)**, 798-799 (1997)
49 Y. Cao, S. Brinkmann-Rengel, N. Nießner, Y. Yang *to be published*
50 J. T. Lai *Synthesis* **2**, 122-124 (1984)
51 EP 0 869 137
52 R. Puts, P. Nicholas, J. Milam, S. Tahilliani, W. Masler, N. Pourahmaday, *Polymer Preprints* **40 (2)**, 323-324 (1999)

# Coordination polymers from kinetically labile copper(I) and silver(I) complexes: True macromolecules or solution aggregates?

Brigitte Lahn, Matthias Rehahn*

German Institute for Polymers and Institute of Macromolecular Chemistry,
Technical University of Darmstadt, Petersenstrasse 22, D-64287 Darmstadt, Germany

SUMMARY: Copper(I) and silver(I) coordination polymers have been prepared via conversion of equimolar amounts of *o*-phenanthroline-based *bis*-bidentate ligand monomers, and $[Cu(CH_3CN)_4]PF_6$ or $AgBF_4$ as the respective metal comonomers. Using NMR spectroscopy, the homogeneous constitution of the diamagnetic products has been proved, and their average chain length has been estimated to be $P_n \geq 20$. Moreover, NMR studies showed the multinuclear complexes to be open (dynamic) solution aggregates when dissolved in solvents that contain coordinating species like acetonitrile or pyridine. When strictly non-coordinating solvents are used, on the other hand, the multinuclear complexes were found to be "true" polymers, *i.e.* macromolecules with a constant number of repeating units per individual chain in time. At very high dilution, finally, transformation of the originally formed chain molecules into cyclic oligomers was observed when coordinating solvents are used, but not in the case of non-coordinating solvents.

## Introduction

High-molecular-weight chain molecules **A** held together by transition-metal complexes, *i.e.* transition-metal coordination polymers, are of considerable interest in macromolecular and supramolecular science because of their electronic, magnetic, optic, catalytic, and solution properties[1,2]. Much effort has been spent, therefore, during recent years to make available a broader variety of such species. Most of the thus developed well-defined polymers are based on thermodynamically stable and *kinetically inert* transition-metal complexes. This selection is reliable because the characterization of the resulting polymers is possible in solution, using essentially the same techniques as are applied for conventional macromolecules. In contrast to this, well-defined coordination polymers from *kinetically labile* transition-metal complexes are still nearly unknown. This is mainly due to the fact that such species decompose simultaneously with their dissolution. Consequently, characterization is possible only in the solid state, and determination of the solution properties is precluded.

Recently, we lined out a number of requirements that we assumed to be essential for obtaining

 CCC 1022-1360/00/$ 17.50+.50/0

readily soluble and thus fully characterizable coordination polymers even from kinetically labile transition-metal complexes[3,4]. The key aspect of our concept was the assumption that coordination polymers decompose exclusively via displacement of the original ligand monomers **L** by coordinating solvent molecules **S** (Scheme 1): even if **S** is not a chelating ligand, it nevertheless can compete successfully for the metal ions because it is present in large number.

*Scheme 1*

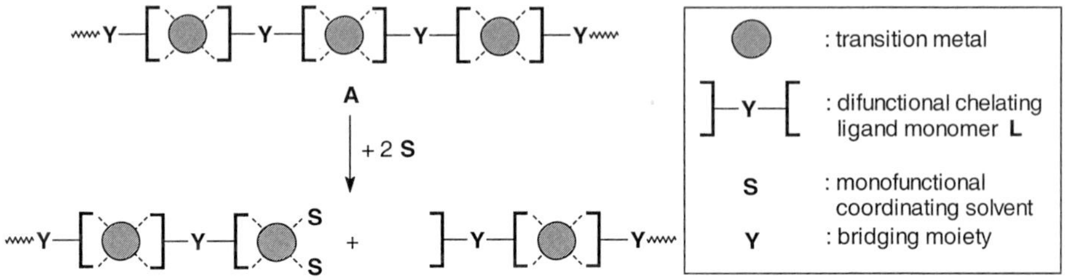


To avoid decomposition according to Scheme 1, coordination polymers **A** from kinetically labile metal complexes must be designed in a way that they are also soluble in strictly non-coordinating solvents. In these media, ligand-exchange should be excluded and hence, the dissolved multinuclear complexes should behave like "true" polymers. In order to probe this concept, we had to take measures to make soluble the coordination polymers — which are polyelectrolytes in general and thus prefer highly polar solvents — also in less polar, non-coordinating media. Attachment of flexible *n*-hexyl side chains was assumed to be one possible method to achieve this goal. Indeed, we were able to verify our hypothesis by means of coordination polymers **1a** and **1b** (Figure 1)[3,4]. Careful analysis of these multinuclear complexes showed them to be well-defined macromolecules in non-coordinating solvents like 1,1,2,2-tetrachloroethane (TCE), but open solution aggregates in the presence of coordinating solvents like acetonitrile or pyridine.

In continuation of this basic work it was of interest to find out to what extent the above strategy can be generalized. One aspect in this regard was the question concerning benefit of the *n*-hexyl side chains of **1**: in addition to their solubilizing effect, they might additionally protect the metal complexes against the attack of coordinating solvent molecules because of their steric demand. A second aim was to find out whether or not entropy forces the kinetically labile multinuclear complexes to decompose in highly diluted solution: cyclooligomerization and/or formation of oligomeric helicates may cause chain degradation and hence prevent full determination of the properties of the polymeric chains in solution. To consider the above aspects, we have prepared the novel copper(I) and silver (I) coordination polymers **2a,b** and **3a,b** (Figure 1).

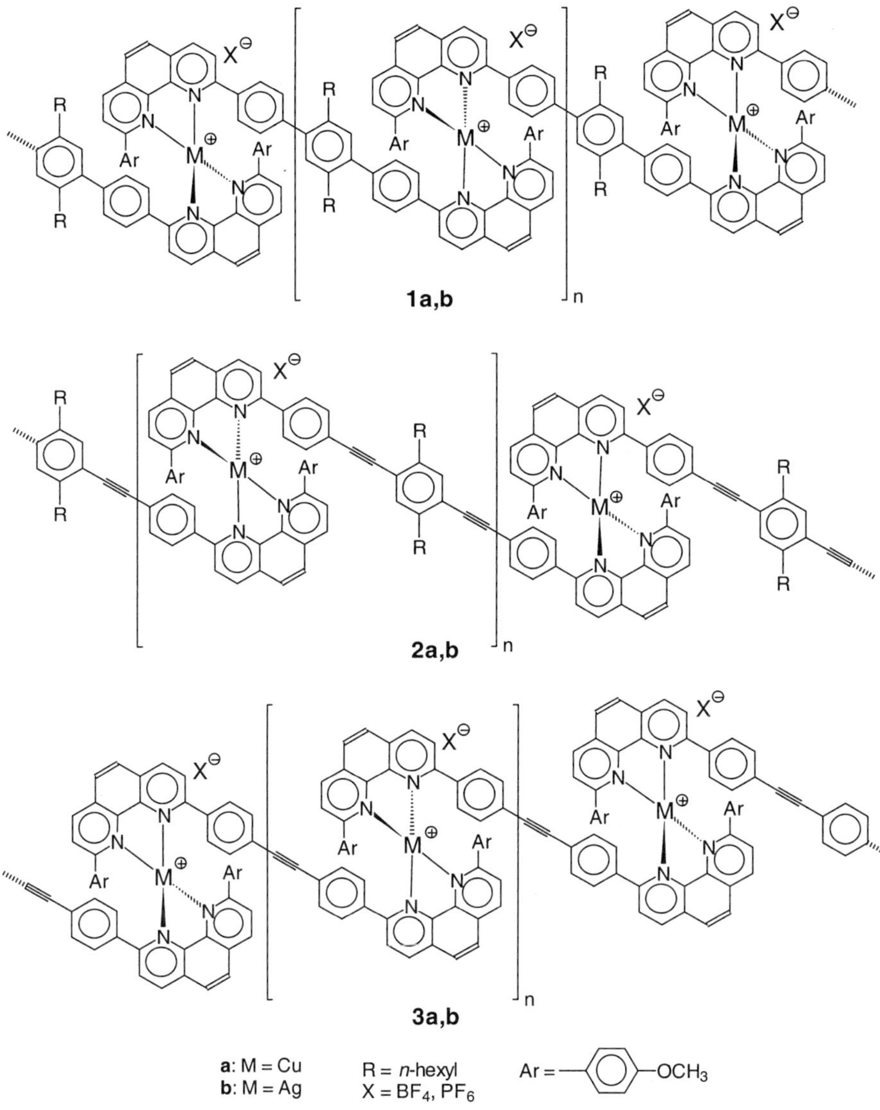


Fig. 1: Coordination polymers **1a,b** - **3a,b**.

In polymers **2**, the *n*-hexyl side chains have a larger distance from the neighboring metal complexes than in polymers **1**, and in polymers **3**, side chains are completely absent. Moreover, because the bridging units **Y** are conformationally rigid in all cases, formation of low-molecular-weight helicates should be rather unlikely. Cyclooligomerization according to Scheme 2, therefore, should be the only way of chain degradation at high dilution except for the destruction of the metal complexes as a whole. Finally, because all other features are quite similar in the three target systems shown in Figure 1, the detailed comparison of the coordination compounds **1** - **3** should provide a deeper insight into the general validity of our concept.

*Scheme 2*

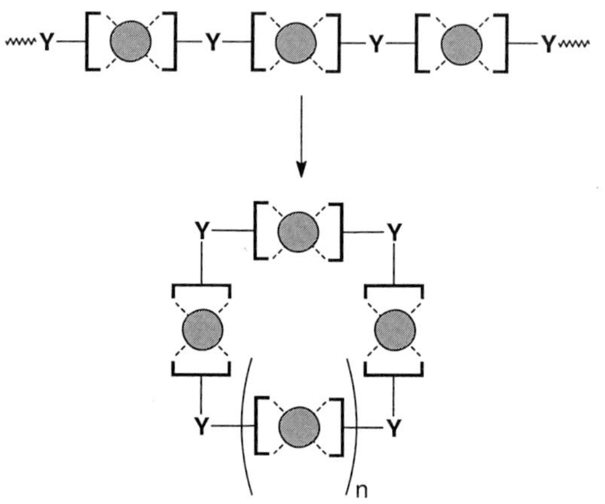


## Monomer synthesis

2(*p*-Bromophenyl)-9(*p*-methoxyphenyl)-*o*-phenanthroline **8** was the central intermediate required for the synthesis of compounds **12** and **15**, the ligand monomers of polymers **2a,b** and **3a,b**. For its synthesis, *o*-phenanthroline **4** was treated with a slight excess of *p*-bromophenyllithium **5** in diethyl ether at - 10 °C. The formed intermediate was dehydrogenated using $MnO_2$, leading to 2-(*p*-bromophenyl)-*o*-phenanthroline **6**[4]. Then, the arylation-rearomatization sequence was repeated with compound **6**, however, using *p*-methoxyphenyl lithium **7** as the arylating agent. From this process, the desired key compound **8** was obtained in approx. 80 % yield.

2,5-Dihexyl-1,4-diethinylbenzene **11** was another starting material required for the synthesis of **12**. It was prepared via Pd-catalyzed condensation of 1,4-dibromo-2,5-dihexylbenzene **9**[5] and trimethylsilylacetylene in the presence of catalytic amounts of $[Pd(PPh_3)_4]$ and CuI, using piperidine as both, solvent and base. The cleavage of the trimethylsilyl protecting groups of **10** was achieved in a mixture of THF, methanol, and sodium hydride. Pure **11** was isolated as a yellow oil (30% yield) which was subsequently treated with a twofold molar excess of **8** under Pd-catalysis. The desired *n*-hexyl substituted ligand monomer **12** was obtained in 60-70% yield.

*Scheme 3*

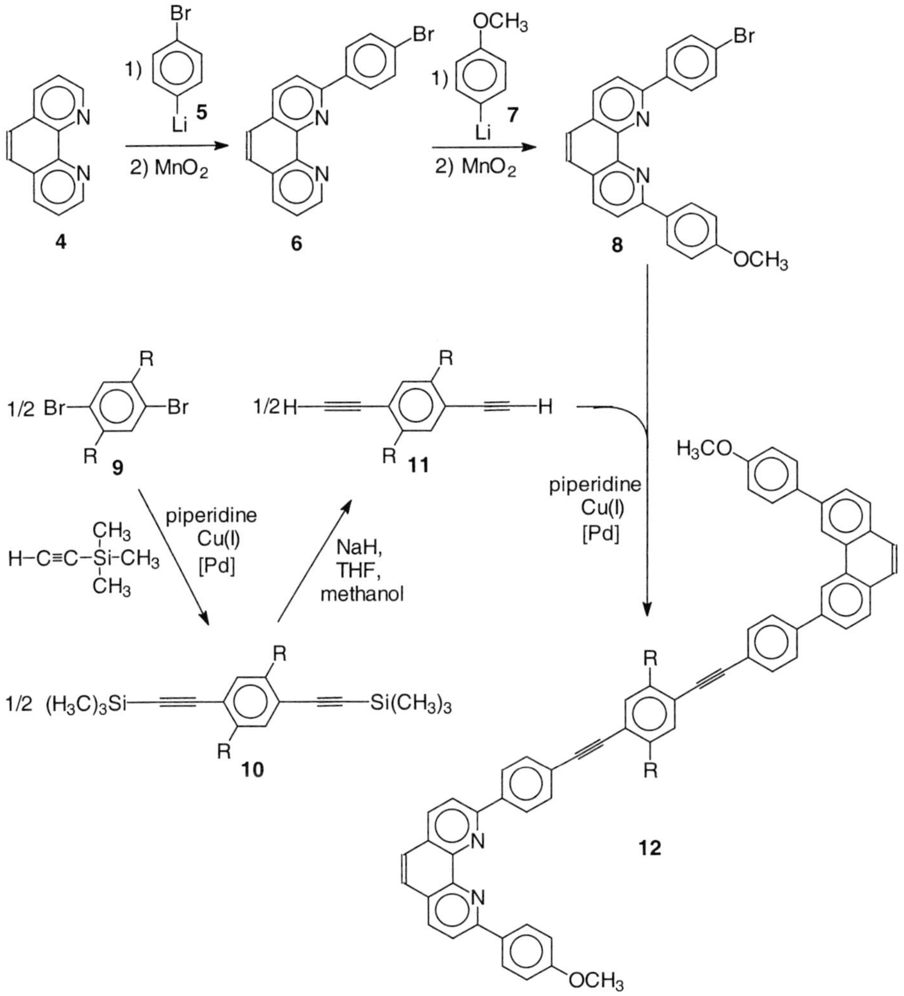

9-(*p*-Methoxyphenyl)-2(*p*-ethinylphenyl)-*o*-phenanthroline **14** was required for the synthesis of ligand monomer **15**. It was prepared in an overall yield of 80% via Pd-catalyzed condensation of equimolar amounts of **8** and trimethylsilylacetylene, followed by cleavage of the trimethylsilyl protecting groups of intermediate **13**. Heating equimolar amounts of **8** and **14** in the presence of $[Pd(PPh_3)_4]$, CuI and piperidine resulted in the pure yellow solid **15** (40% yield).

*Scheme 4*

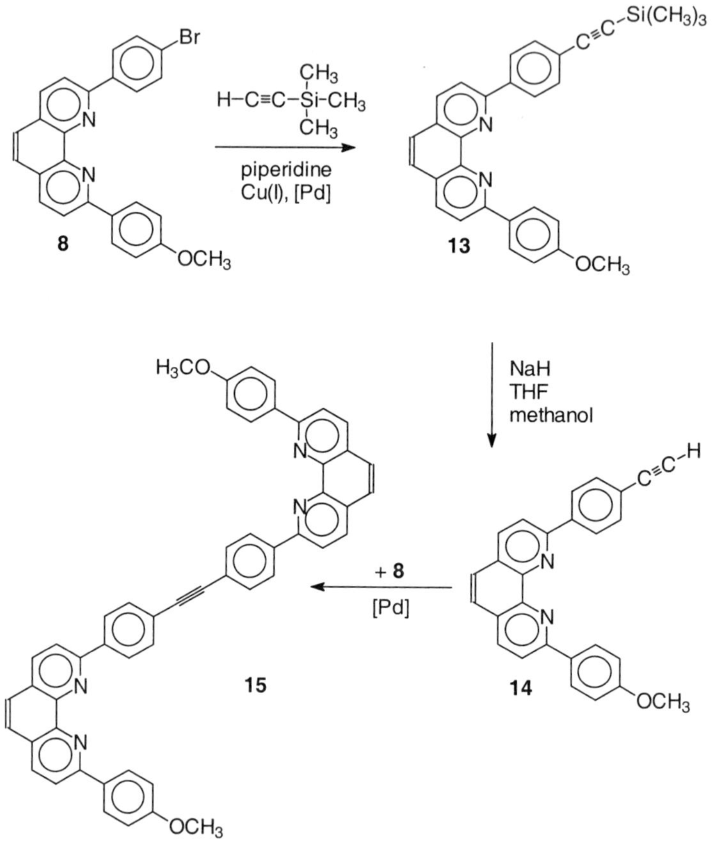


The molecular constitution of ligand monomers **12** and **15** as well as of all isolated intermediates was verified using $^1H$ and $^{13}C$ NMR spectroscopy and elemental analysis. The purity of the ligand monomers **12** and **15** was estimated to be > 98% ($^1H$ NMR).

## Polymer synthesis

Most of the subsequent conversions of ligand monomers **12** and **15** into the copper(I) and silver(I) coordination compounds **2a,b** and **3a,b**, respectively, were carried out in NMR tubes, using the same procedure as described recently for the synthesis of **1a,b**[4]. We started our studies by preparing the copper(I) system **2a** (Scheme 5): 1.0 equiv. (approx. 50 mg) of **12**, 0.8 equiv. of $[Cu(CH_3CN)_4]PF_6$ **16**, and approx. 1 mL of an 4:1 (v/v) mixture of TCE-$d_2$ and acetonitrile-$d_3$ were weighed into an NMR tube[a]: The mixture was slightly heated and subsequently, a $^1H$ NMR spectrum was recorded of the resulting homogeneous solution.

*Scheme 5*

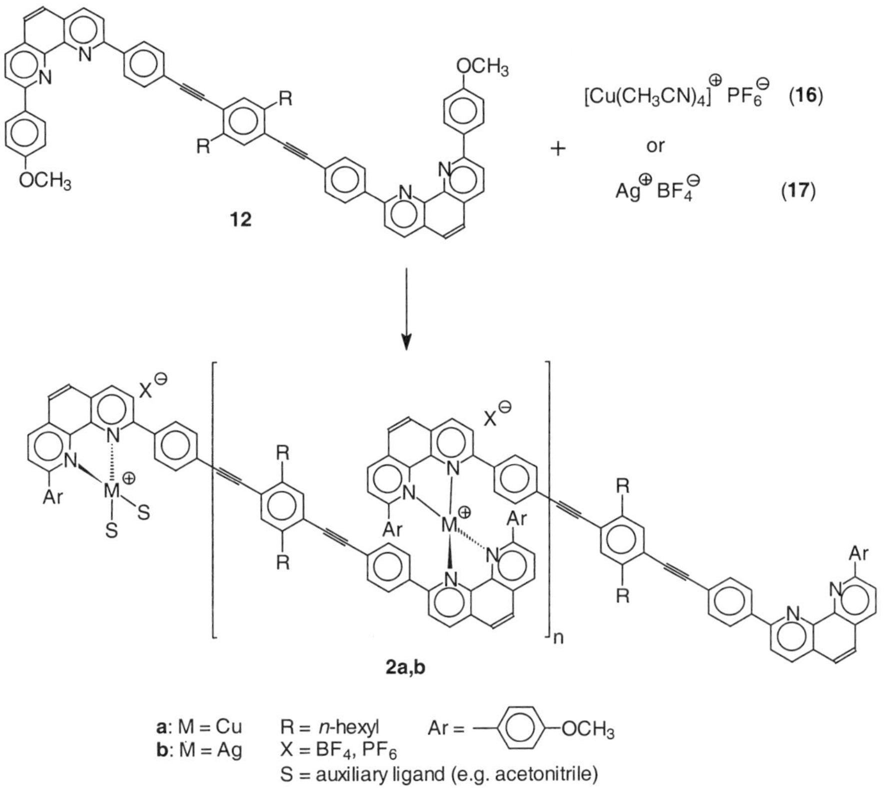


[a] This solvent mixture was found to be well-appropriate for these conversions: TCE is a good solvent for the ligand monomers as well as for the resulting coordination polymers. Acetonitrile, on the other hand, readily dissolves the metal monomers like **16** and **17**. Moreover, it catalyzes the formation of the desired $[M(phen)_2]^+$ motifs.

Afterwards, further small quantities of **16** were added to the solution in steps of approx. 0.05 equiv. Each time, a $^{1}H$ NMR spectrum was recorded after complete conversion. As an example, Figure 2B shows the spectrum of an oligomeric complex **2a** prepared from 1.0 equiv. of **12** and 0.8 equiv of **16**. Figure 2A, on the other hand, displays the spectrum of ligand monomer **12**. Comparison of these two spectra makes evident that the desired (pseudo)tetrahedral $[Cu(phen)_2]^+$ motifs and hence the multinuclear complexes **2a** were formed: the characteristic upfield shift of most of the intense absorptions in spectrum B represents a clear proof of the desired catenation of ligand monomers **12** via copper(I) complexes[b]. In addition to the intense absorptions of the inner-chain repeating units of **2a** there are observable signals of minor intensity (assigned by asterisks in Figure 2B). We assign these absorptions to the chain-termini of **2a** which were expected because of the 1.0 : 0.8 molar ratio of the comonomers used here. Moreover, the ratio of the signal intensities of inner-chain repeating units and well-defined endgroups [in particular those of the $OCH_3$ protons at $\delta \approx 4.0$ ppm (terminal methoxy groups) and $\delta \approx 3.5$ ppm (inner-chain methoxy groups)] resulted in an average degree of polymerization of $P_n \approx 6$-8. This estimate is in good agreement with the comonomer ratio applied here.

When the 1 : 1 equivalence of the comonomers was finally adjusted, all the small absorptions disappear completely, and only the intense signals of the inner-chain repeating units remain. The spectrum does not change any more even if the formed product **2a** is isolated via precipitation in *n*-hexane and redissolved in pure TCE-$d_2$ (Figure 2C): the lack of endgroup absorptions and of absorptions that might correspond to structural defects proves both, the homogeneous constitution and the considerable chain length of **2a**. This conclusion was additionally supported by the $^{13}C$ NMR spectra: there as well, a full signal assignment could be done. To conclude, **2a** is stable also in the solid state and has — (*i*) due to the lack of endgroup absorptions, (*ii*) considering the limits of accuracy of the $^{1}H$ NMR method, and (*iii*) assuming that cyclic oligomers are not present here[c] — a value of $P_n \geq 20$.

A quite different behavior was observed when the corresponding silver(I) polymer **2b** was prepared. In order to illustrate this difference, Figure 2D shows the $^{1}H$ NMR spectrum of a reaction mixture obtained via conversion of 1.0 equiv. of ligand monomer **12** and 0.8 equiv. of $AgBF_4$ **17** in TCE-$d_2$ / acetonitrile-$d_3$ (4:1, v/v). The intense absorptions of the inner-chain

---

[b] This upfield shift is due to magnetic shielding effects caused by the ring currents of the aromatic units of the neighboring ligand monomers: in polymers like **2a**, the succeding ligand monomers are spatially arranged in a way that their protons are placed "above" (or "below") the *o*-phenanthroline rings of the neighboring ligand monomers.

repeating units of **2b** are readily detectable and support the formation of the desired multinuclear complex **2b**. But surprisingly, no endgroup absorptions can be observed in this spectrum, despite of the fact that **12** is present in a 20% excess. This is in clear contrast to the corresponding spectrum of **2a** (Fig. 2B) and might be due to a very fast ligand-exchange occurring in the silver(I) complexes, smearing out the endgroup absorptions. To verify this hypothesis, the oligomeric product **2b** was precipitated in *n*-hexane, carefully dried, redissolved in pure TCE-$d_2$, and studied again by NMR. This was done because the rate of ligand exchange should strongly depend on the concentration of acetonitrile. Supposed our initial hypothesis is valid, i.e. the coordinating co-solvent acetonitrile supports the ligand-exchange, endgroup absorptions should appear as soon as acetonitrile is absent. In fact, the missing absorptions are observable in the latter $^1$H NMR spectrum (signals assigned by asterisks in Figure 2E). Now, by means of these absorptions, the precise 1:1 equivalence of the comonomers could be adjusted easily, and high-molecular-weight samples of **2b** could be obtained which do not show any endgroup signals even in pure TCE-$d_2$ (Figure 2F). $^1$H and $^{13}$C NMR spectra moreover proved the constitutional homogeneity of the formed product. All together, it is important to emphasize that disappearence of endgroup absorptions for oligomers was not observed in the case of the silver(I) complexes **1b**. This is a strong evidence for the higher lability of the complexes incorporated in coordination compounds **2**. The less efficient shielding of the metal complexes of **2** due to their larger distance from the *n*-hexyl side chains might be responsible for this finding.

Now, we broadened our studies to systems **3** which do not bear aliphatic side chains at all. Here as well, we started the experiments by means of the copper(I) species, using TCE / acetonitrile as the solvent system. The $^1$H NMR spectra recorded from reaction mixtures containing an excess of either **15** or **16** were dominated by the intense absorptions of the inner-chain repeating units of **3a**. Nevertheless, there were also detectable the absorptions of the chain termini. Therefore, it was possible to adjust the 1:1 equivalence of the comonomers **15** and **16** using $^1$H NMR spectroscopy very precisely. Finally, the high chain length as well as the homogeneous constitution of **3a** could be proved by means of $^{13}$C NMR spectroscopy as well. To conclude, the solution properties of **3a** are quite similar to those of **2a**. The only significant difference is the much higher solubility of **2a** due to its solubilizing *n*-hexyl side chains.

---

[c] Cyclooligomerization would cause disappearence of the endgroup absorptions as well.

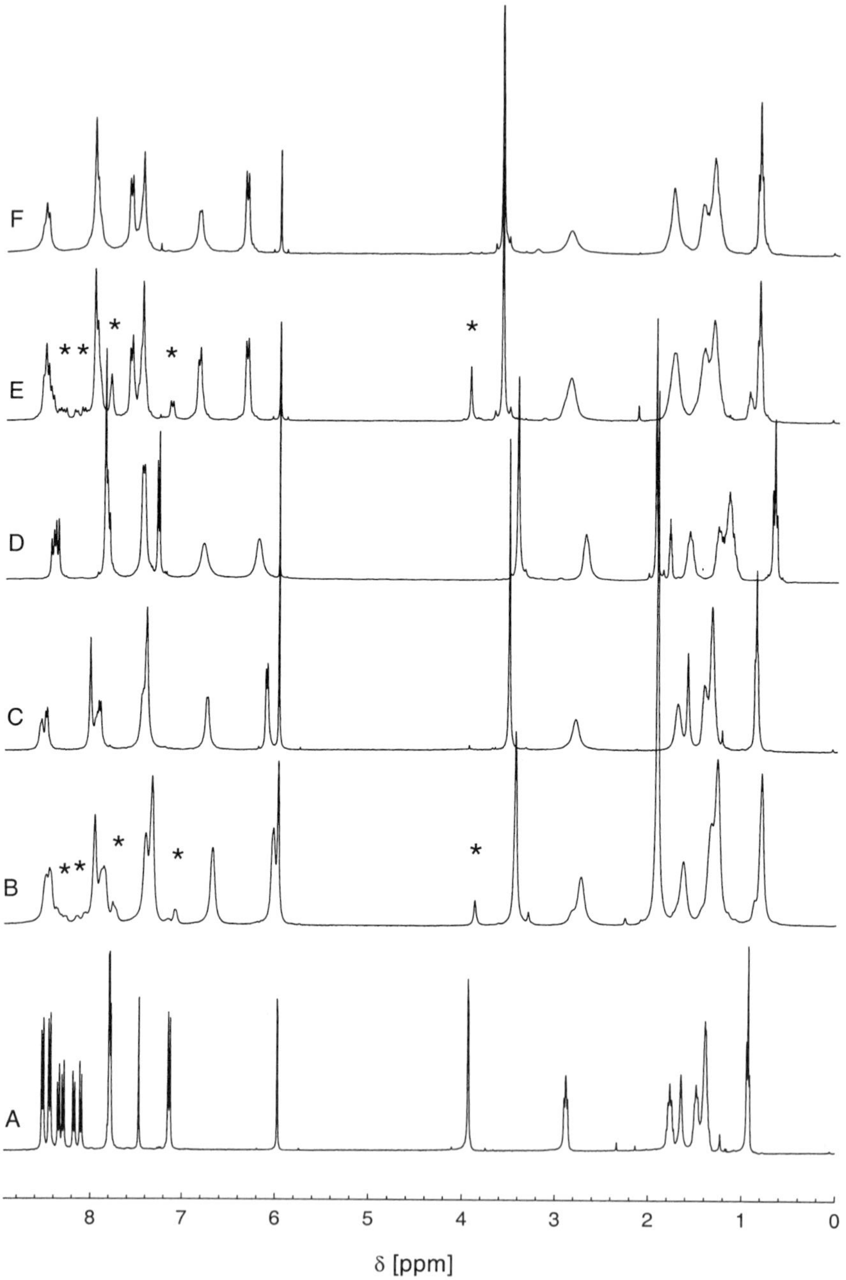


Fig. 2: $^{1}$H NMR spectra of (A) ligand monomer **12** (TCE-$d_2$), (B) oligomeric complex **2a** (TCE-$d_2$/acetonitrile-$d_3$), (C) polymeric complex **2a** (TCE-$d_2$), (D) oligomeric complex **2b** (TCE-$d_2$/acetonitrile-$d_3$), (E) oligomeric complex **2b** (TCE-$d_2$), and (F) polymeric complex **2b** (TCE-$d_2$); see text for details.

Based on these results, the solution properties of the silver(I) compound **3b** were expected to be quite similar to those of **2b**. In full agreement with that, intense $^{1}H$ NMR absorptions were found for the inner-chain repeating units of **3b**, but endgroup absorptions could not be detected in the solvent system TCE / acetonitrile despite of an excess of **15** or **17** (20 % excess of **15** in the case of the spectrum shown in Figure 3B). After isolation of the formed oligomers and redissolution in pure TCE, however, the latter absorptions were detected as well (Figure 3C). Now, the precise 1:1 equivalence of the comonomers could be be adjusted (Figure 3D), and the homogeneous constitution as well as the considerable chain length of **3b** was proved using $^{1}H$ and $^{13}C$ NMR spectroscopy. To summarize, the above experiments show that — under otherwise identical conditions — the kinetic lability of the multinuclear silver(I) complexes **2b** and **3b** is higher than that of the corresponding copper(I) systems **2a** and **3a**. Moreover, it is evident that the lability of the polymeric complexes **2a** and **3a** as well as of **2b** and **3b** is quite similar. In other words, the aliphatic side chains of **2** are too far away from the metal centers and thus cannot cause a significant stabilization as was found for polymers **1**. Nevertheless, the *n*-hexyl side chains are of considerable benefit in polymers **2** because they increase the solubility considerably as one can see from the comparison with polymers **3**.

The quantities of **2** and **3** available via NMR titration are too small for a detailed analysis of their solution properties. Therefore, large-scale syntheses were performed, taking advantage of the elution time that the reaction mixture needs to pass a capillary to adjust the 1:1 equivalence of the comonomers. A three necked flask was equipped with a capillary viscosimeter, and we started the syntheses using 1.0 : 0.8 mixtures of the respective ligand- and metal monomers. After complete conversion, the elution time of the reaction mixture was determined. Subsequently, further quantities of metal monomer were added in steps of approx. 0.05 equiv. After each addition, the elution time was determined. A drastic increase was found until the 1:1 equivalence was reached. Then, if further metal monomer was added, the elution times dropped down again. In order to come back to the maximum elution time, the required quantity of ligand monomer was added. As an example, Figure 4 displays a characteristic plot recorded during the synthesis of **2a**. After isolation of the formed products as reddish-brown [copper(I) systems] or yellow [silver(I) systems] fibrous materials, it was possible to show by means of $^{1}H$ and $^{13}C$ NMR spectroscopy that really high-molecular-weight chain molecules of homogeneous constitution were formed under these conditions as well.

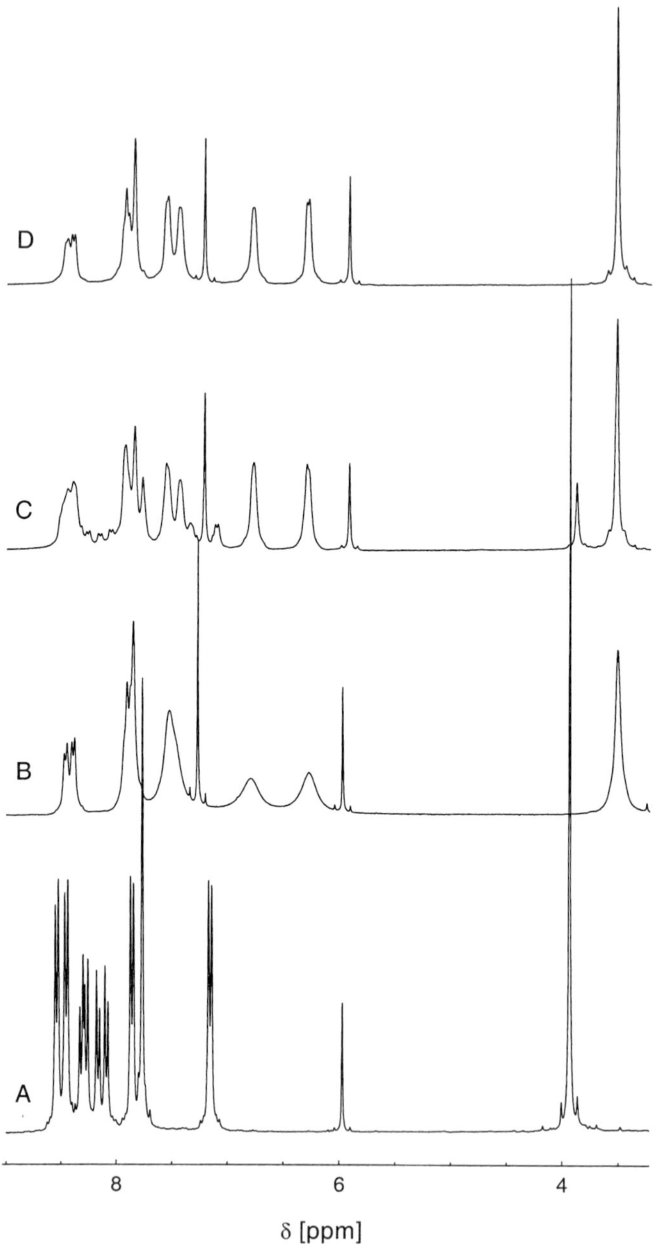


Fig. 3: $^{1}H$ NMR spectra of (A) ligand monomer **15** (TCE-$d_2$), (B) oligomeric complex **3b** (TCE-$d_2$/acetonitrile-$d_3$), (C) oligomeric complex **3b** (TCE-$d_2$), and (D) polymeric complex **3b** (TCE-$d_2$); see text for details.

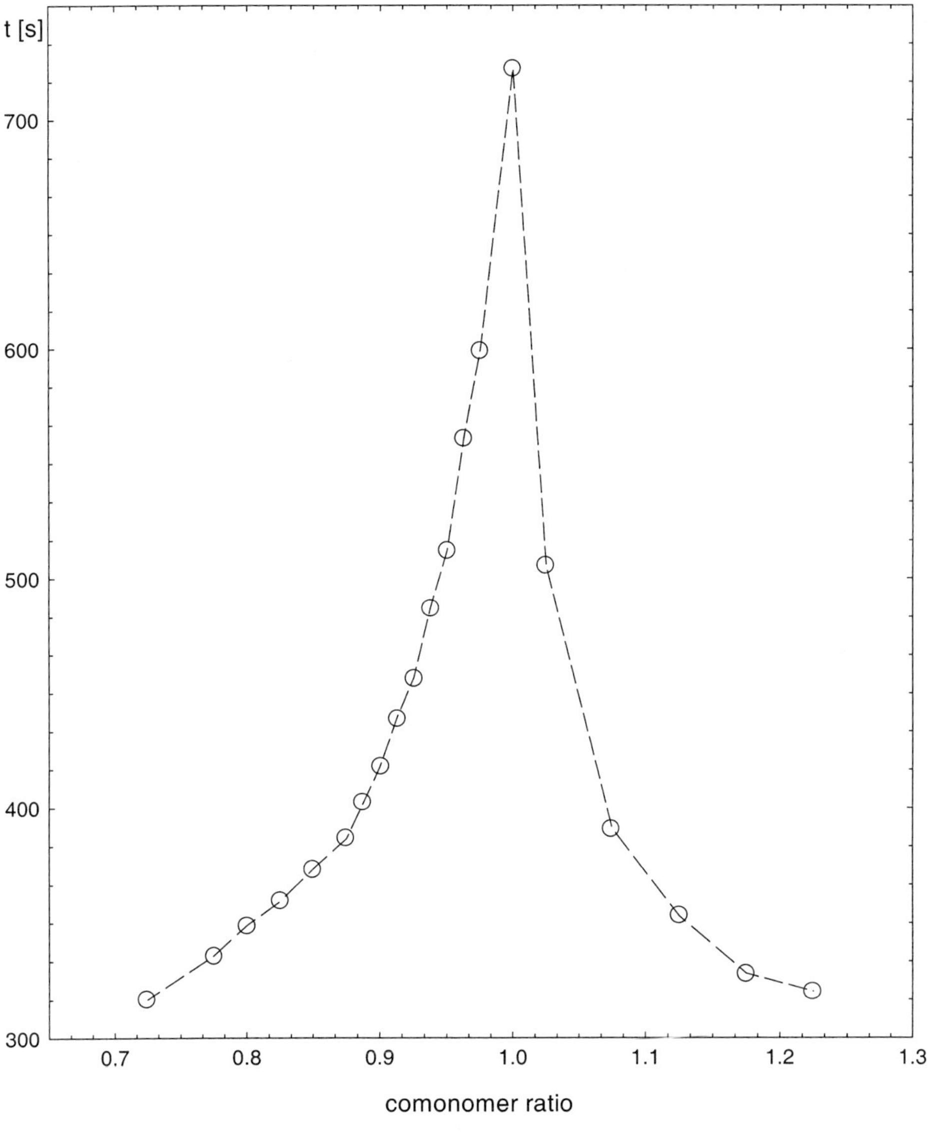


Fig. 4: Elution time of the reaction mixture vs. molar ratio of the comonomers **12** and **16**; the experiment was performed in acetone/acetonitrile. See text for details.

## Properties in solution

The investigations described so far showed the multinuclear complexes **2a,b** and **3a,b** to undergo perpetual ligand exchange when dissolved in a coordinating solvent. Hence, under these conditions, the chains are dynamic (open) solution aggregates. On the other hand, ligand exchange is dramatically slowed down when no auxiliary ligands are available in the solution. So far, however, we did not show whether or not the ligand exchange is completely stopped in the latter media. Therefore, we carried out two further series of NMR studies in order to show under which conditions the multinuclear complexes are real macromolecules. First, a series of hybridization experiments was performed using various deuterated solvent systems (Scheme 6).

*Scheme 6*

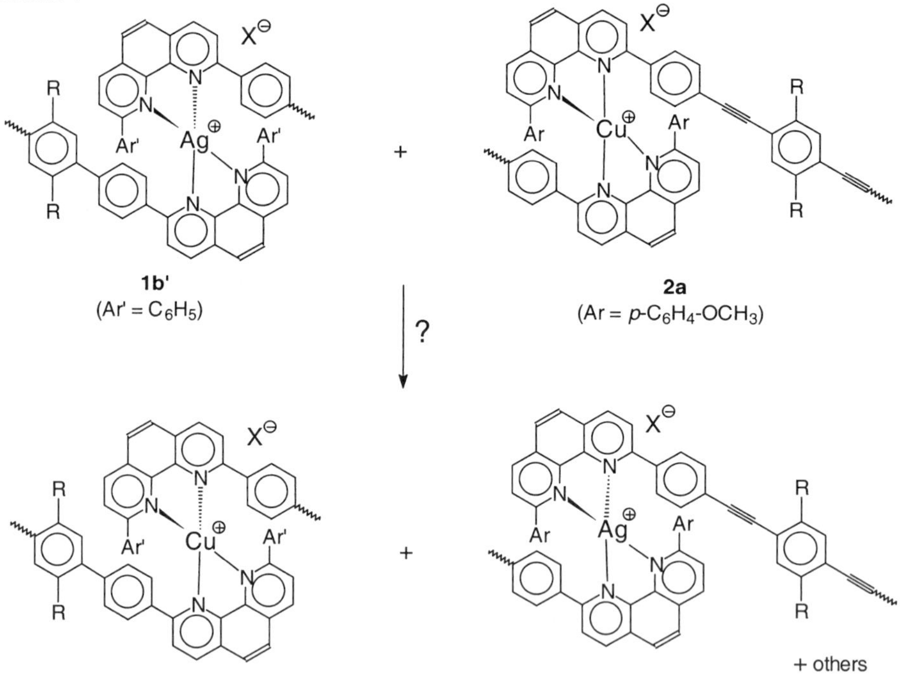


**1b'** (Ar' = $C_6H_5$)

**2a** (Ar = $p$-$C_6H_4$-$OCH_3$)

**1a'** (Ar' = $C_6H_5$)

**2b** (Ar = $p$-$C_6H_4$-$OCH_3$)

R = $n$-hexyl, X = $BF_4$, $PF_6$

In these experiments, we took advantage of the fact that the terminal methoxy groups of the ligand monomers absorb at slightly different chemical shifts depending on whether they are attached to a copper(I) or a silver(I) complex. In practice, the solution of a copper(I) system (**2a** or **3a**) was mixed with a solution of the silver(I) polymer **1b'**. The latter does not bear methoxy groups at its phenyl substituents ("Ar" in Scheme 6). Vice versa, solutions of the silver(I) systems (**2b** or **3b**) were mixed with solutions of the respective copper(I) polymer **1a'**. After mixing, NMR spectra were recorded of the resulting solution at regular intervals in order to follow an eventual ligand-exchange according to Scheme 6. This is easily possible by analyzing in particular the singlet of the methoxy protons at $\delta \approx 3.5$ ppm. As an example, Figure 5 shows the spectra recorded in the course of such an experiment performed according to Scheme 6. Spectra A and B are those of the pure starting polymers **1b'** and **2a**, respectively, recorded in pure TCE-$d_2$. Immediately after mixing the solutions (spectrum C) as well as after 3 days (spectrum D), only one single absorption is observable at $\delta \approx 3.5$ ppm. This result proves the practically complete lack of ligand exchange in the absence of coordinating species. If, however, small amounts of a coordinating solvent such as acetonitrile (or water) are added, the ligand-exchange starts immediately. This process is clearly detectable by the appearence of a second singlet at $\delta \approx 3.7$ ppm as well as of further absorptions in the aromatic region (spectrum E).

In the second series of NMR experiments, we studied the behavior of the multinuclear complexes at very high dilution. By means of these investigations, we wanted to find out in which solvents the chain molecules either remain unchanged or decompose into small, presumably cyclic oligomers upon diluting. The concentration of the investigated solutions varied from $c_P \approx 20$ g·L$^{-1}$ down to $c_P \approx 0.1$ g·L$^{-1}$, and non-coordinating as well as coordinating solvents were tested. In full agreement with the above results, no change was observed in the NMR spectra even at the lowest concentration when TCE was the solvent: only the well-known absorptions of the inner-chain repeating units were observed in all spectra (see Figure 6 for **2a**), proving the complete stop of ligand exchange in this non-coordinating solvent. When, on the other hand, the experiments were carried out in the presence of acetonitrile, or in slightly humid acetone, decomposition of thc chain molecules into cyclic oligomers was observed: new absoptions appear in particular in the aromatic region, clearly not identical to those of the chain termini whose chemical shifts are well-known from the linear oligomeric systems discussed above. As an example, Figure 7 displays a series of $^1$H NMR spectra of **2a**, recorded in acetone-$d_6$ at various concentrations.

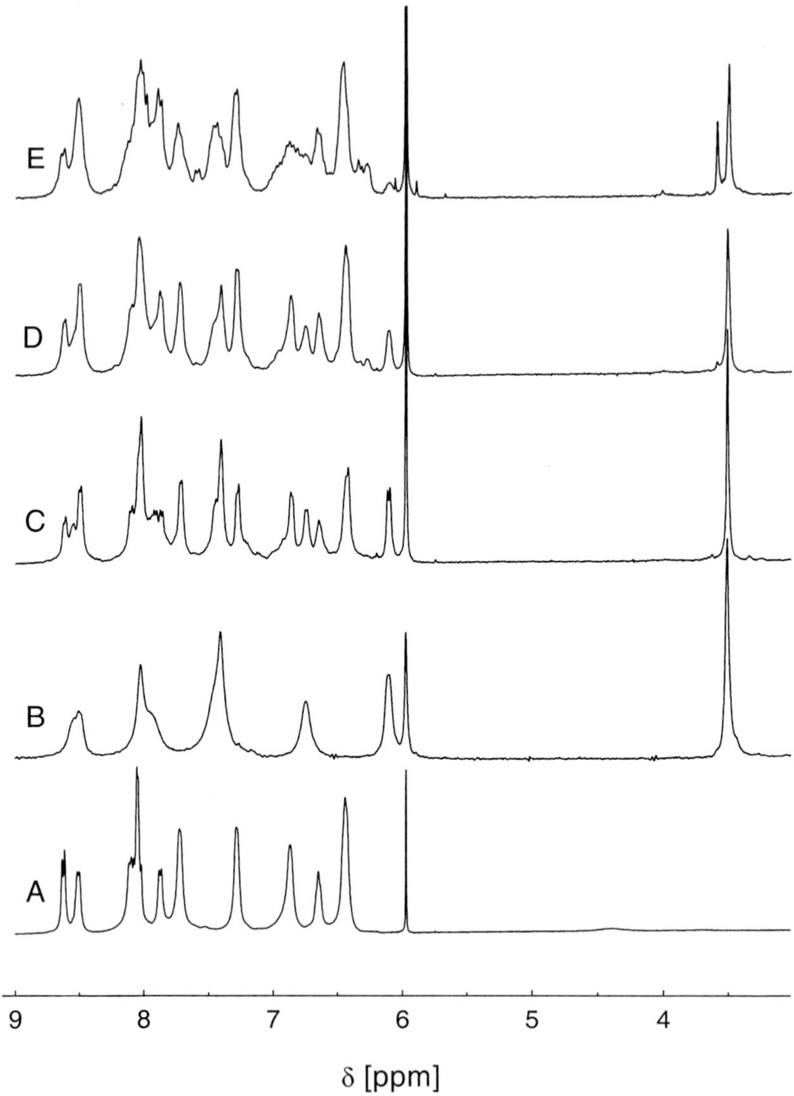


Fig. 5: $^{1}$H NMR spectra of (A) polymeric complex **1b'** (TCE-$d_2$), (B) polymeric complex **2a** (TCE-$d_2$), (C) the mixture thereof after 5 min. (TCE-$d_2$), (D) the mixture thereof after 3 d (TCE-$d_2$), and (E) the mixture thereof after addition of acetonitrile; see text for details.

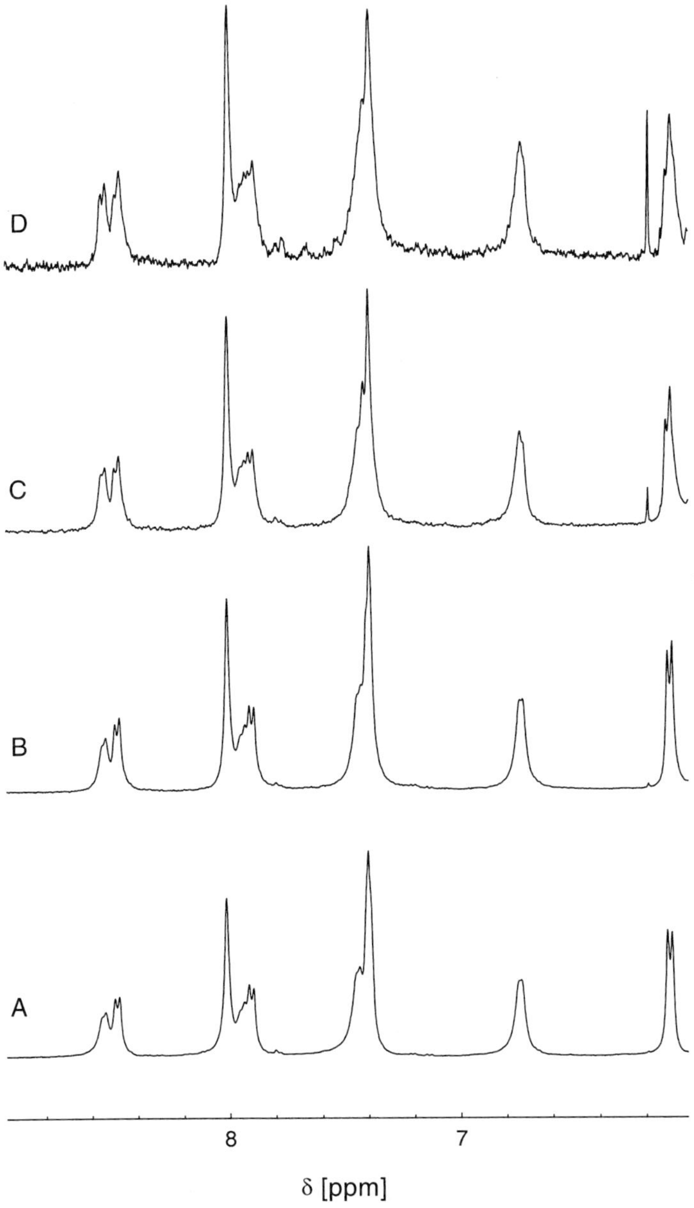


Fig. 6: $^1$H NMR spectra of **2a** recorded at a concentration of $c_P$ = (A) 10 g·L$^{-1}$, (B) 5 g·L$^{-1}$, (C) 1 g·L$^{-1}$, and (D) 0.1 g·L$^{-1}$ (TCE-$d_2$). The sharp singlet at δ = 6.2 ppm is due to the solvent; see text for details.

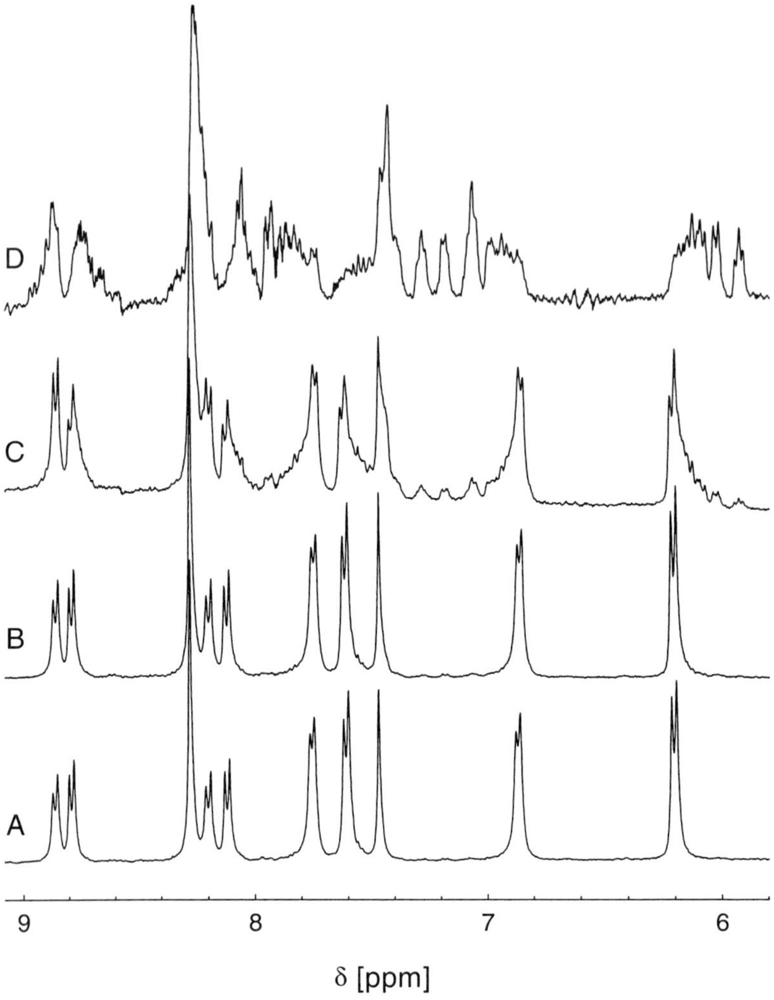


Fig 7: $^1$H NMR spectra of **2a** recorded at a concentration of $c_P$ = (A) 10 g·L$^{-1}$, (B) 5 g·L$^{-1}$, (C) 1 g·L$^{-1}$, and (D) 0.1 g·L$^{-1}$ (acetone-$d_6$); see text for details.

While the solution leading to spectrum A ($c_P \approx 10$ g·L$^{-1}$) does not show any evidence of cyclization, there are observable new, well-defined absorptions at $c_P = 5$ g·L$^{-1}$ (spectrum B). In particular the sequences of signals at $\delta \approx 6$ ppm and at $\delta \approx 7.2$ ppm are very significant in this regard. Their intensities increase further with decreasing $c_P$ (spectra C, D). Since none of these absorptions is due to open chain termini, it is reliable to assume that the chain molecules change into well-defined cyclic oligomers (helicates seem to be rather unlikely due to the rodlike shape of the ligand monomers). This interpretation is supported by the above-mentioned series of new signals at $\delta \approx 6$ ppm and at $\delta \approx 7.2$ ppm which we believe correspond to cyclic oligomers of different ring sizes ("n" in Scheme 2). The obvious change of the linear macromolecules into low-molecular-weight species is additionally reconfirmed by viscosimetry which provides clear evidence of a systematic chain degradation upon dilution.

## Conclusions and Outlook

Well-defined transition-metal coordination polymers **2a,b** and **3a,b** were prepared via conversion of ligand monomers **12** or **15** with metal monomers **16** or **17**. The homogeneous constitution and the high molecular weights ($P_{n,NMR} \geq 20$) of the products could be proved using $^{1}$H and $^{13}$C NMR spectroscopy. The polymers can be prepared in the gramm scale, are easily to handle, and are stable for months when stored as dry solids in the dark. The rate of ligand exchange is high in the presence of coordinating solvents and very low in non-coordinating solvents. This allows us to consider the systems as real polymers in the latter media but as open solution aggregates in the former. Moreover, ligand exchange is much more efficient in the silver(I) systems than in the analogous copper(I) species. Also, upon comparing polymers **2** and **3** with systems **1** recently reported, it is possible to conclude that the *n*-hexyl side chains do not only solubilize the coordination polymers but can additionally stabilize the metal complexes — provided their distance is not too large. A further key result of the present study is that it is possible to "trap" the macromolecules as long chains even at high dilution by using a strictly non-coordinating solvent. However, as soon as only traces of a coordinating solvent are added to these solutions, the chains degrade immediately, changing to presumably cyclic oligomers. Presently, we prepare more efficiently stabilized copper(I) and silver(I) coordination polymers to make feasible a full analysis of the solution properties also at very low concentration.

## Acknowledgement

The authors are grateful to Prof. M. Ballauff, Karlsruhe, for his support of the work. Financial support by the Deutsche Forschungsgemeinschaft is gratefully acknowledged.

## References

1. M. Rehahn, *Acta Polymer.* **49**, 201 (1998).
2. I. Manners, *Angew. Chem.* **108**, 1713 (1996).
3. U. Velten, M. Rehahn, *J. Chem. Soc., Chem. Commun.* 2639 (1996).
4. U. Velten, B. Lahn, M. Rehahn, *Macromol. Chem. Phys.* **198**, 2789 (1997).
5. M. Rehahn, A.-D. Schlüter, W. J. Feast, *Synthesis* 386 (1988).

*Macromol. Symp.* ***2001****, 163, 177-187*


# Functionalized Oligomers and Copolymers with Metal Complexing Segments: A Simple and High Yield Entry towards 2,2':6',2''-Terpyridine Monofunctionalized Telechelics

Ulrich S. Schubert,[a,b]* and Christian Eschbaumer[a]

a) Lehrstuhl für Makromolekulare Stoffe, Technische Universität München, Lichtenbergstr. 4, D-85747 Garching, Germany
b) *Center for NanoScience*, Ludwig-Maximilians-Universität München, Geschwister-Scholl-Platz 1, D-80593 München, Germany

SUMMARY: 4'-Chloro-2,2':6',2''-terpyridine was reacted in a high yield Williamson type reaction with α-hydroxy-ω-carboxy-functionalized poly(ethylenoxide) to obtain monoterpyridine terminated telechelics. The completeness of the functionalization was proven by NMR, GPC and MALDI-TOF-MS investigations. Addition of transition metal ions resulted in the formation of the corresponding octahedral terpyridine metal complexes and resulted in the formation of metallo-supramolecular dimers.

## Introduction

The blending of supramolecular with macromolecular systems is a major aim in today's research (see, e.g.[1,2]). In particular, metallo-supramolecular polymer composites could provide novel organic-inorganic hybrid materials with a wide range of interesting potential properties, e.g. concerning smart materials, intelligent glues or electro- and photoactive compounds. There are several different entries into this field. Besides the final complexation of special block copolymers and micelles with metal binding sites or compartments[3], the construction of composites utilizing non-covalent interactions is of special interest[4]. For practical applications, supramolecular segments with an availability in usable quantities and with recognition-directed self-assembly features are necessary. This requires a careful selection of suitable building blocks. Very promising systems come from the area of *N*-heterocyclic ligands, which have been known for nearly a century as very effective and stable complexation agents for transition metal ions[5]. In particular 2,2':6',2''-terpyridine compounds,

 CCC 1022-1360/00/$ 17.50+.50/0

which spontaneously form stable octahedral metal complexes with a wide range of transition metal ions (such as cobalt(II), copper(II), manganese(II), iron(II), iron(III), chromium(III), nickel(II) or ruthenium(II)) have been studied extensively[6)], due to their interesting electronic and redox properties[7)]. However, the difficulty in obtaining a straightforward and high yield preparation of terpyridine-containing oligomers and polymers has always been a major drawback in this direction. Usually, special terpyridine ligands were functionalized with hydroxy, amino or carboxy end groups and then reacted in a polycondensation type reaction with suitable monomers, oligomers or polymers (see e.g.[8)]). Recently we developed a different approach in this direction utilizing very simple and meanwhile commercially available 4'-functionalized terpyridine ligands as new starting materials (for first preliminary results, see[9)]). As first example we used α-hydroxy-ω-carboxy-functionalized poly(ethylen-oxide) oligomers to prepare monofunctionalized terpyridine-containing systems (Fig. 1). Addition of suitable transition metal ions to these oligomers resulted in mono-complexes or in a dimerisation and thus in a doubling of the molecular weight. In addition, the carboxyl end group allows the attachment onto surfaces, e.g. via silane chemistry, onto glass or $Si_3N_4$ substrates. Extension of this strategy leads towards oligomers and polymers with two or more terpyridine end groups.

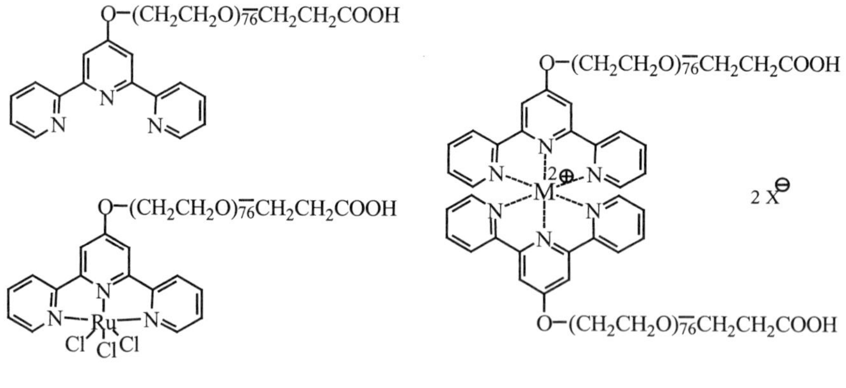


Fig. 1: Schematic representation of the synthesized monofunctionalized oligomers and the corresponding metal-containing systems (M = Co, Ru, Zn).

# Experimental

## Materials and Characterization

Materials were obtained from commercial suppliers and used without further purification. Anhydrous dimethylformamide (DMF) was purchased from Aldrich Chemical Company. $^{1}$H and $^{13}$C NMR spectra were recorded at 300 and 75 MHz, respectively, on a Bruker AC 300 spectrometer. The chemical shifts were calibrated to the residual solvent peak or TMS. Gel permeation chromatography (GPC) analysis was performed on a Waters Liquid Chromatograph system using Shodex GPC K-802S columns, Waters Differential Refractometer 410 and Waters UV Absorption Detector 486 with chloroform as eluent. Calibration was conducted with polystyrene standards. UV/VIS measurements were recorded using a Varian Cary 3 UV/VIS spectrometer. MALDI-TOF-mass spectra were measured with Bruker Biflex 3 mass spectrometer (matrix dithranol, solvent acetone).

## Synthesis:

**Preparation of the 4'-chloro-2,2':6',2"-terpyridine 1:** Synthesized according to[10]. Mp. 152°C; $^{1}$H-NMR ($CHCl_3$, 300 MHz): 7.35 (dd), 7.87 (dd), 8.49 (s), 8.60 (d), 8.71 (d).

**Preparation of the α-carboxyl-ω-(2,2':6',2"-terpyrid-4'-yl)-poly(ethylenoxide) 2:** To a suspension of KOH (0.900 g, 16.07 mmol) in DMSO (10 ml) α-carboxy-ω-hydroxy-poly-(ethylenoxide) (0.100 g, 27 μmol) was added. After stirring 1 h at 60°C 4'-chloroterpyridine (0.015 g, 54 μmol) was added. The mixture was stirred at 60°C for 24 h. The heater was removed and the mixture was poured into cold water (50 ml). The solution was extracted with chloroform (3 × 100 ml). The combined organic phases were dried over $Na_2SO_4$ and the solvent removed *in vacuo*. The residue was solved in THF and dropwise added to diethyl ether. α-Carboxyl-ω-(2,2':6',2"-terpyrid-4'-yl)-poly(ethylenoxide) precipitates as a white solid yielding 95 mg (26 μmol, 96%). $^{1}$H-NMR ($CDCl_3$, 300 MHz): δ (ppm) 2.60 (t, $J$ = 6.1 Hz, 4 H, $H_{PEG}$), 3.40 (t, $J$ = 4.8 Hz, 4 H, $H_{PEG}$), 3.64 (m, 410 H, $H_{PEG}$), 4.42 (m, 3 H, $H_{PEG}$), 7.35 (m, 2 H, H-5,5"), 7.87 (m, 2 H, H-3',5'), 8.07 (m, 2 H, H-3,3"), 8.66 (m, 4 H, H-6,6"). MALDI-TOF-MS: $\overline{M}_n$ : 3635; $\overline{M}_w$ : 3659, PDI: 1.01. GPC ($CHCl_3$, PS-standard): $\overline{M}_n$ : 6202; $\overline{M}_w$ : 6808, PDI: 1.09. UV/VIS (acetonitrile): λ(ε) = 239 nm (26100), 276 nm (23900). Anal. Calc. for $C_{170}H_{319}N_3O_{79}$: calcd.: C, 55.65; H, 8.76; N, 1.15; found: C, 55.71; H, 8.27; N, 1.05.

**Preparation of the α-carboxyl-ω-(2,2:6',2''-terpyrid-4'-yl-ruthenium(III)chlorid)-poly-(ethylenoxide) 3:** To a solution of $RuCl_3 \times 3\ H_2O$ (4.0 mg, 15 μmol) in ethanol (4 ml) α-carboxyl-ω-(2,2':6,2''-terpyrid-4'-yl)-poly(ethylenoxide) was added. After stirring 4h at 60°C the heater was removed. The mixture was cooled to ambient temperature when brown precipitate was formed. After 4 h at –78°C the precipitat was separated and washed with cold ethanol yielding 35 mg (9.03 μmol, 66%) α-carboxyl-ω-(2,2':6,2''-terpyrid-4'-yl-ruthenium(III)chlorid)-poly(ethylenoxide). $^1$H-NMR (300MHz, $CDCl_3$): $\delta$ (ppm) 3.64 - 3.77 (m, $H_{PEG}$); MALDI-TOF-MS: $\overline{M}_n$ : 3361; $\overline{M}_w$ : 3420, PDI: 1.02; GPC ($CHCl_3$, PS-standard): $\overline{M}_n$ : 4706; $\overline{M}_w$ : 6956, PDI: 1.48. UV/VIS (methanol): λ(ε) = 275 nm (26118), 303 nm (15304), 391 nm (5532).

**Preparation of the cobalt(II) polymer ion complex 4:** To a solution of α-carboxyl-ω-(4'-terpyridyl)-poly(ethylenoxide) (0.478 mg, ) **2** in methanol (5 ml) a solution of $Co(CH_3COO)_2 \times 4\ H_2O$ (1.118 mg in 10 ml methanol) was added dropwise. The formation of the polymer ion complex was monitored by UV/VIS-spectroscopy. UV/VIS (methanol): λ(ε) = 308 (14573). MALDI-TOF-MS: $\overline{M}_n$ : 7525; $\overline{M}_w$ : 7568, PDI: 1.01.

**Preparation of the zink(II) polymer ion complex 5:** To a solution of $ZnCl_2$ (10 mg, 73 μmol) a solution of α-carboxyl-ω-(4'-terpyridyl)-poly(ethylenoxide) **2** (20 mg, 5.45 μmol) in methanol was added dropwise. The successful formation of the single sided polymer ion complex was detected by MALDI-TOF mass spectrometry. MALDI-TOF-MS: $\overline{M}_n$ : 3738; $\overline{M}_w$ : 3792, PDI: 1.01.

**Preparation of the model complex *bis*[4'-oxytetramethylenoxy$^t$butyl-(2,2':6',2''-terpyridine)]cobalt(II) hexafluorophosphat 6:** Synthesized according to[11]. UV/VIS (methanol): λ(ε) = 307 (27500). MALDI-TOF-MS: 814 ($M^+$ - 2 $PF_6^-$).

## Results and Discussion

The combination of *N*-heterocyclic ligands and other metal complexing segments with oligomers and polymers has attracted much interest during the last decade. Such materials could offer a new entrance to organic-inorganic hybrid composites. Besides the applications in bulk material the use in thin films and on surfaces is of particular interest. Systems with one metal complexing unit could be e.g. tethered to substrates or used for the formation of supramolecular dimers (Fig. 2).

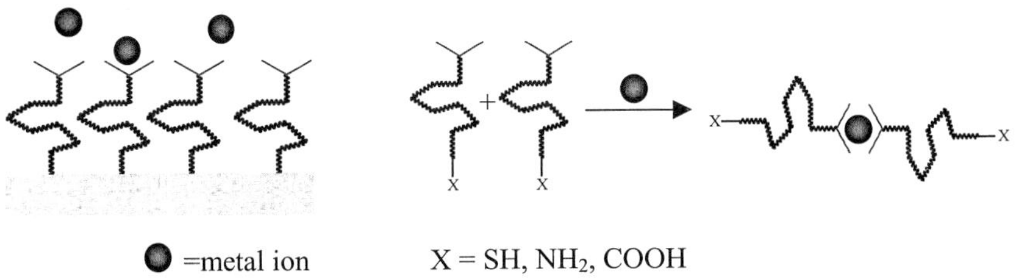


Fig. 2: Schematic representation of polymers with one metal complexing unit: Bound to surfaces (left) or utilized for the formation of dimers (right).

In most cases bipyridine or terpyridine ligands have been used, due to the outstanding complexing abilities of these units. Recently, we have demonstrated the incorporation of 5,5''-*bis*(hydroxymethyl)terpyridine segments into the backbone of a block copolymer and the successful crosslinking in presence of transition metal ions like cobalt(II). However, all these approaches suffer from the complicated and time consuming synthetic accessibility of the utilized functionalized *N*-heterocyclic segments. Therefore we developed a new synthetic strategy towards terpyridine containing systems based on a different starting material: 4'-chloro-2,2':6',2''-terpyridine **1** (Fig. 3).

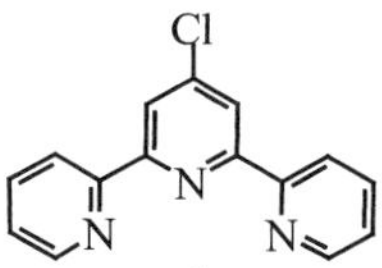


**1**

Fig. 3: Schematic representation of the central terpyridine building block **1**.

This very simple and now commercially available unit (Aldrich) can be synthesized in multihundred gram quantities utilizing a modified Kröhnke synthesis[10)]. The ligand can be used directly for the functionalization of hydroxy terminated oligomers and polymers. As a first example we reacted a commercially available α-carboxy-ω-hydroxy functionalized prepolymer **2** with 4'-chloro-2,2':6',2''-terpyridine **1** in a Williamson type ether synthesis in the presence of potassium hydroxide in anhydrous DMSO. The desired α-carboxyl-ω-(4'-terpyridyl)-poly(ethylenoxide) **3** with both a complexing and a carboxyl end group could be isolated by precipitation in 88% yield (Fig. 4).

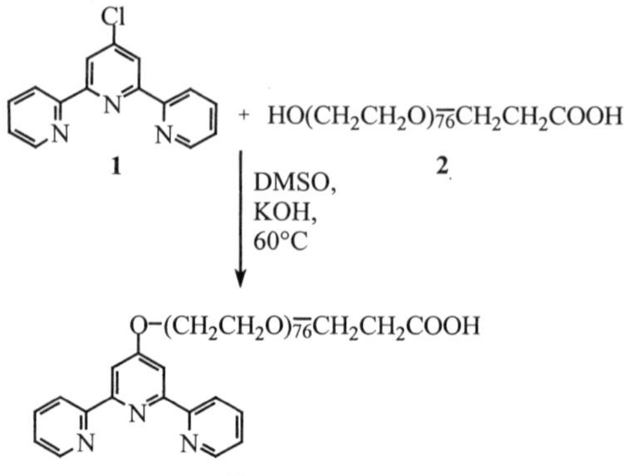


3

Fig. 4: Synthesis of α,ω-functionalized metal complexing polymers **3**.

The complete monofunctionalization of the poly(ethylenoxide) with the terpyridine moiety could be proven by various techniques. Fig. 5 compares the $^1$H-NMR-spectra of the basic telechele and the terpyridine containing product. The left inset in the product spectrum (b) shows the signals of the terpyridine with the characteristic shift of the protons 3 and 5 after exchange of the chloro by an oxygen group in position 4 (see also[12)]).

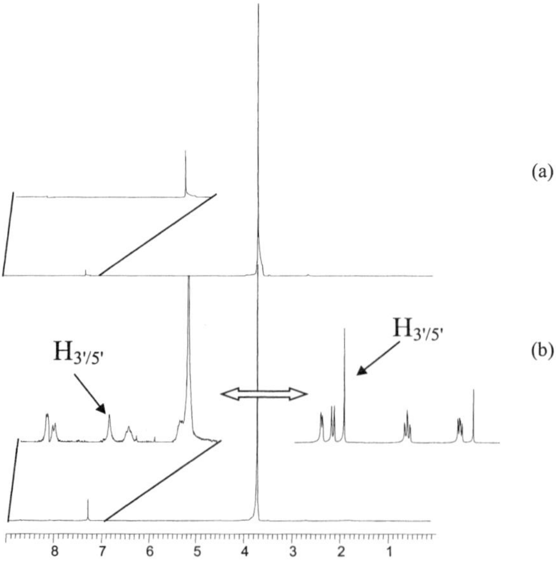


Fig. 5: $^1$H-NMR spectra: (a) α-carboxyl-ω-hydroxy-poly(ethylenoxide) **2**; (b) α-carboxyl-ω-(4'-terpyridyl)-poly(ethylenoxide) **3**; the right inset shows the aromatic region of the 4'-chloro-terpyridine **1** spectrum.

Utilizing GPC the incorporation of the terpyridine moiety can be easily seen by applying a UV/VIS detector (254 nm) due to the different absorption properties of the poly(ethylenoxide) and the terpyridine unit (Fig. 6).

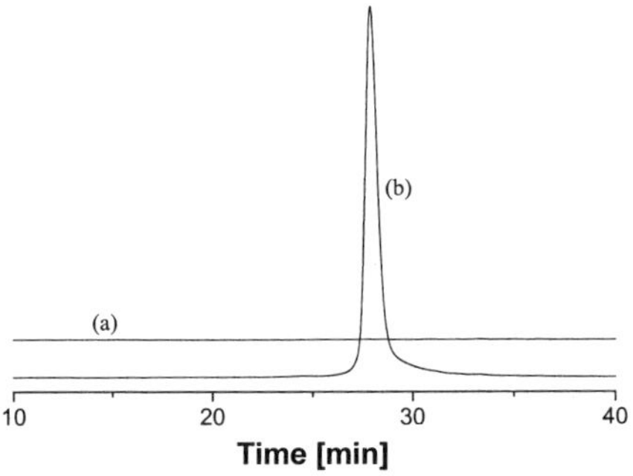


Fig. 6: GPC results of (a) α-carboxyl-ω-hydroxy-poly(ethylenoxide) **2** and (b) α-carboxyl-ω-(4'-terpyridyl)-poly(ethylenoxide) **3** utilizing a UV/VIS detector (254 nm).

However, this does not provide a quantitative proof of a complete functionalization. The question can be addressed by using MALDI-TOF mass spectrometry (Fig. 7). The difference in molecular weights between prepolymer **2** and the product **3** correlates with the complete reaction of one terpyridine unit for each prepolymer.

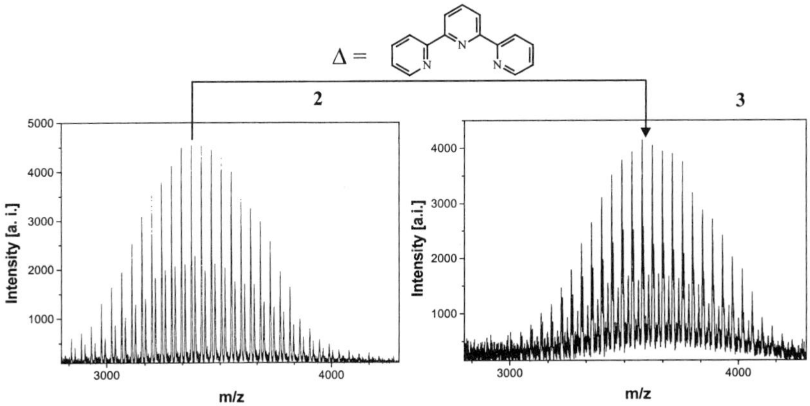


Fig. 7: MALDI-TOF-MS: Left) α-Carboxyl-ω-hydroxy-poly(ethylenoxide) **2**; right) α-carboxyl-ω-(4'-terpyridyl)-poly(ethylenoxide) **3**.

The successful functionalization can also be demonstrated by the addition of transition metal ions such as ruthenium(III), which can selectively form 1:1 complexes with terpyridine ligands[13)] (Fig. 8). This allows the construction of unsymmetrical dimers.

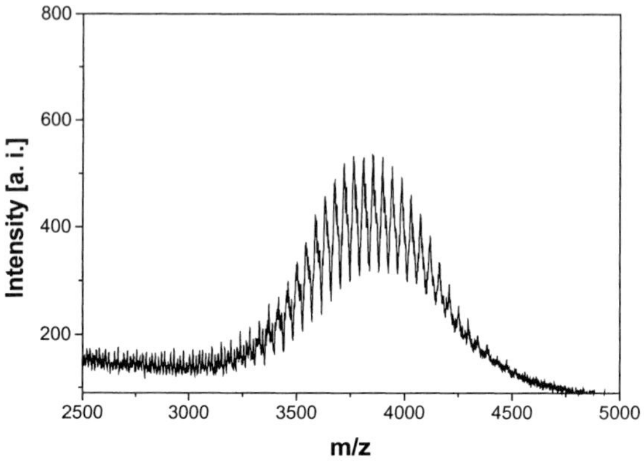


Fig. 8: MALDI-TOF-MS of the α-carboxyl-ω-(4'-terpyridyl)-poly(ethylenoxide) **4** ruthenium(III) complex.

Addition of other transition metal ions, such as cobalt(II) or zink(II) leads, to a spontaneous self-assembly of two terpyridine units and thus to a dimerisation (Fig. 9). The formation of the complexes can be reversed, e.g. by changing pH or temperature, or by applying electrochemical changes[14)].

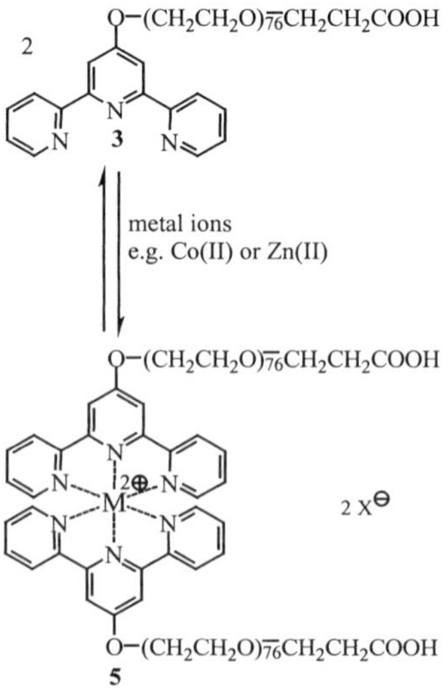


Fig. 9: Schematic representation of the dimerization of oligomer **3** by addition of transition metal ions, such as cobalt(II).

The complexation can easily be monitored by different techniques. Fig. 10 shows the change of the UV/VIS absorption behavior following addition of cobalt(II) acetate to the terpyridine-containing polymer. A redshift of the $\pi$-$\pi^*$-band can be observed which is typical for the formation of terpyridine transition metal complexes.

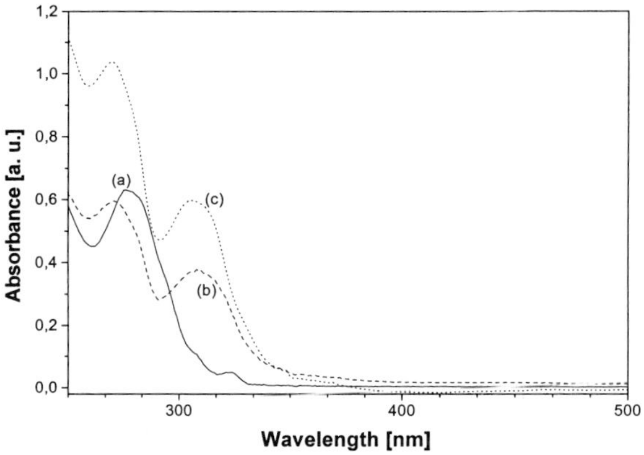


Fig. 10: UV/VIS spectra of (a) the uncomplexed polymer **3**, (b) the polymer ion complex **5** in methanol, (c = 458 µg/5ml), and (c) the terpyridine-cobalt-model complex **6**.

Again, MALDI-TOF mass spectrometry revealed to be very useful for the characterization of such metallo-supramolecular assemblies and polymers: As shown in Fig. 11, the molecular weight of the polymer has doubled after addition of metal ions.

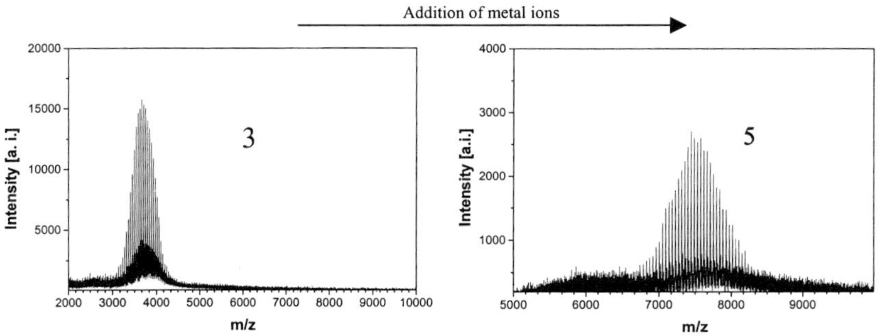


Fig. 11: MALDI-TOF MS of the terpyridine containing polymer **3** (left) and the polymer-ion complex **5** (right).

Extension of this methology can lead to systems functionalized with two or more terpyridine units and therefore to a non-covalent polymerization (in principle a polyaddition reaction) or crosslinking (Fig. 12).

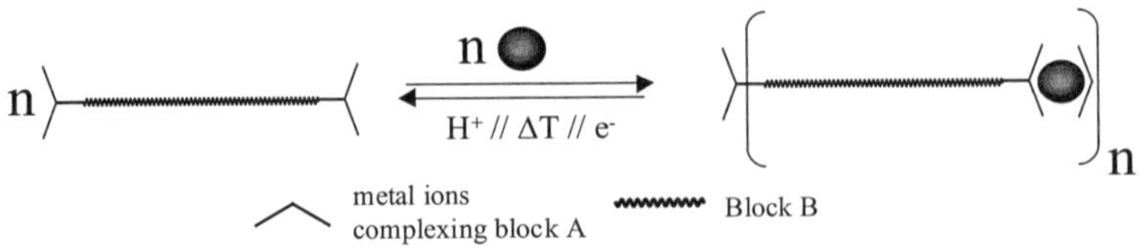


Fig. 12. Schematic representation of a potential non-covalent polymerization.

## Conclusions

In this paper we showed a simple and high yield synthesis of a new class of terpyridine monofunctionalized metal complexing telechelics. The complete functionalization could be proven by various techniques, including NMR spectroscopy, MALDI-TOF mass spectrometry and GPC. The complexation of the polymer by addition of transition metal ions results in the case of ruthenium(III) ions in 1:1 complexes and in the case of metal ions such as cobalt(II) in the formation of non-covalent dimers. Further studies will concentrate on the investigation of this materials and the preparation of *bis*functional and multifunctional building blocks.

## Acknowledgements

The research was supported by the *Bayerisches Staatsministerium für Unterricht, Kultus, Wissenschaft und* Kunst (Bayerischer Habilitations-Förderpreis for USS), the *Fonds der Chemischen Industrie* and the *Deutsche Forschungsgemeinschaft (DFG, SFB 266)*. We thank O. Hien for the synthesis of some of the 4'-chloroterpyridine and Prof. Dr.-Ing. O. Nuyken for his support.

## References

1. J.-M. Lehn, *Supramolecular Chemistry - Concepts and Perspectives*, VCH, Weinheim, Germany (1995)
2. D. S. Lawrence, T. Jiang, M. Levett, *Chem. Rev.* 2229 (1995)
3. Y. Chujo, K. Sada, T. Saegusa, *Macromolecules* **26**, 6315 (1993); P. Ng, X. Gong, W. T. Wong, W. K. Chan, *Macromol. Rapid Commun.* **18**, 1009 (1997); J. J. S. Lamba, C. L. Fraser, *J. Am. Chem. Soc.* **119**, 1801 (1997); J. P. Spatz, A. Roescher, M. Möller, *Adv. Mater.* **8**, 337 (1996); J. P. Spatz, S. Mößmer, M. Möller, *Chem. Eur. J.* **2**, 1552 (1996); J. P. Spatz, S. Mößmer, M. Möller, *Angew. Chem., Int. Ed. Engl.* **35**, 1510 (1996); M. Antonietti, S. Henke, A. Thünemann, *Adv. Mater.* **8**, 41 (1996); M. Antonietti, C. Göltner, *Angew. Chem.* **109**, 944 (1997); *Angew. Chem., Int. Ed. Engl.* **36**, 910 (1997);

U. S. Schubert, C. Eschbaumer, C. H. Weidl, *Polymeric Materials Science and Engineering (ACS), Proceedings* **80**, 191 (1999)

4. See, e.g.: C. D. Eisenbach, U. S. Schubert, *Macromolecules* **26**, 7372 (1993); U. Velten, M. Rehahn, *Macromol. Chem. Phys.* **199**, 127 (1998)
5. E. C. Constable, *Metals and Ligand Reactivity*, VCH, Weinheim, Germany (1996); G. Wilkinson, R. D. Gillard, J. A. McCleverty (Eds.), *Comprehensive Coordination Chemistry*, Pergamon Press, Oxford, **2** (1987)
6. B. N. Figgis, E. S. Kucharski, A. H. White, *Aust. J. Chem.* **36**, 1527 (1983); R. Bhula, D. C. Weatherburn, *Aust. J. Chem.* **44**, 303 (1991)
7. W. R. McWhinnie, J. D. Miller, *Adv. Inorg. Chem. Radiochem.* **12**, 135 (1969); E. C. Constable, *Adv. Inorg. Chem. Radiochem.* **30**, 69 (1986); E. C. Constable, A. M. W. C. Thompson, *New. J. Chem.* **16**, 855 (1992)
8. U. S. Schubert, C. Eschbaumer, C. H. Weidl, *Design. Monom. Polym.* **2**, 185 (1999)
9. U. S. Schubert, C. Eschbaumer, *Polym. Preprints* **40(2)**, 1070 (1999)
10. K. T. Potts, D. Kanwar, *J. Org. Chem.* **56**, 4815 (1991); E. C. Constable, M. Ward, *J. Chem. Soc., Dalton Trans.* 1405 (1990); O. Hien, *Diploma Thesis*, TU München (1998)
11. U. S. Schubert, O. Hien, C. Eschbaumer, P. Andres, in preparation.
12. G. R. Newkome, E. He, *J. Mater. Chem.* **7**, 1237 (1997)
13. B. P. Sullivan, J. M. Calvert, T. J. Meyer, *Inorg. Chem.* **19**, 1404 (1980)
14. R. Farina, R. Hogg, R. G. Wilkins, *Inorg. Chem.* **7**, 170 (1968)

*Macromol. Symp.* **2001,** *163, 189-212* 

# Interfacial design in fiber reinforced polymers

Edith Mäder*, Elena Pisanova

Institute of Polymer Research, Hohe Str. 6, 01069 Dresden, Germany

SUMMARY: Different interphases have been created with different film formers and coupling agents on glass and carbon single filaments embedded in thermoplastic and thermosetting matrices. Three micromechanical procedures (pull-out test, fragmentation test and a version of microbond test in which crack propagation was continuously monitored by optical microscopy) were used to measure fiber/matrix interfacial bond strengh. The effect of interphase microstructure, transcrystallinity as well as matrix molecular weight on the measured bond strength, failure mode, and local properties of the interphase was examined. The possibilities of controlling bond strength between fiber surface and polymer matrix are discussed.

## Introduction

Mechanical properties of the fiber reinforced composites are strongly influenced by the properties of the interfacial region between the fiber and the matrix. The interphases comprise a significant part of the matrix phase even in the moderately filled composites [1], so specific properties of the boundary layers should be taken into account by modeling and designing advanced materials. Recently new non-destructive methods have been developed to characterize physical properties of the interphases. It was demonstrated by means of atomic force microscopy [2, 3], dynamic mechanical analysis [4, 5], scanning electron microscopy [6, 7], inverse gas chromatography [8, 9] that the properties of the polymer in the vicinity of the fiber can change drastically. The reasons for this may be restricted segmental mobility, transcrystallinity, selective adsorption of a component of polymeric composition, changed curing, specific interaction between the fiber surface and the matrix polymer. These processes lead to creation of an interphase giving gradient of the matrix properties at the interface. On the other hand, interphases can be designed between fiber and matrix to control composite properties, first of all, stress transfer through the interface.

The theoretical consideration has shown that the stress distribution at the fiber-matrix interface can be considerably changed depending on interphase elastic modulus and thickness [4, 10]. That means, interfacial design should be based on foregoing analysis of all possible interactions for a given polymer-fiber pair. It is important that the goals of the design may be opposite for different classes of composites. Indeed, for many thermoplastic polymers, especially when they are the

 CCC 1022-1360/00/$ 17.50+.50/0

non-polar, the main issue is an increase in interfacial adhesion. On the contrary, most thermosetting polymers possess very good adhesion to carbon and glass fiber surface, what results in brittle fracture and low impact resistance. In this case interphases should be created, for example, by introducing a ductile or flexible material between the fiber and the matrix [4].

The interfacial design can be performed when a composite is being formed, or in part before that (matrix modification, fiber surface treatment). But the effectiveness of these can be proved only after the composite formation. Two research directions are of great importance on this stage: first, direct measurements of the interphase properties, and second, micromechanical experiments aimed at examining the ability of the boundary layer of stress transfer. Recently developed micromechanical techniques that provide simultaneous observation of crack initiation and propagation [11–14] deliver necessary information on the interfacial strength as well as on the failure location. This makes possible a separate analysis of the adhesion and friction contributions into measured bond strength. Such an information is a necessary basis for further target-oriented interfacial design.

In this work, various approaches were used to control the structure and properties of the interphases between glass and carbon fibers and polymer matrices of different nature: polar and non-polar thermoplastics as well as thermosetting resins.

## Experimental

### *Materials*

E-glass fibers with a diameter of 12 μm and PAN-based carbon fibers with a diameter of 7 μm were used for the experiments. As matrix material, we used granules of thermoplastic polymers — polypropylene homopolymer (PP) and maleic anhydride grafted (PPM), high density polyethylene homopolymer (PE) and maleic anhydride/styrene grafted (PEMS), poly(vinylidene fluoride) (PVF), poly(acrylonitrile butadiene styrene) (ABS), polyamide 6 (PA), polycarbonate (PC), polystyrene (PS) and polyphenylene sulphide (PPS). As thermosetting polymers, vinyl ester resin and DGEBA-based epoxy resin were taken.

***Sublayer formation***

— *From solution*: different sizing agents were applied from an aqueous solution or polymer dispersion within the continuous spinning process, immediately after cooling glass fibers down. Following model sizings were used: γ-aminopropyltriethoxysilane (APS), methacrylsilane (MAS), APS/polyurethane dispersion (APS+PU), MAS/unsaturated polyester resin dispersion (MAS+UPE), polyvinylalcohol (PVA) and polyvinylacetate dispersion (PVAc).

— *By electrodeposition*: sublayers on carbon fibers were obtained by depositing powdered polymer onto continuous fibers in the chamber of electrostatic fluidized bed coater [15].

***Micromechanical tests***

— *Pull-out test*: this test was carried out using a pull-out apparatus which allows high precision fiber displacement and force measurement as well as data management [16]. Single fibers were embedded in matrices in a separate micro-oven at a heating/cooling rate of about 50 °C/min. The pull-out test was performed at a crosshead displacement rate of $1.2 \cdot 10^{-2}$ mm/min. From each force-displacement curve, the maximum force $F_{max}$, and the embedded length, $l_e$, were determined, and the apparent shear strength $\tau_{app}$, for each specimen was calculated from the equation

$$\tau_{app} = \frac{F_{\max}}{\pi d l_e},$$

where $d$ is fiber diameter.

— *Fragmentation test*: single-fiber composites in the form of a dumbbell, with the fiber running along the specimen axis, were prepared as described elsewhere [15]. The specimen stretching was performed on a tension testing machine at a strain rate of 0.2 mm/min until a neck of maximum possible length was reached. After that specimen was examined in an optical microscope. The interfacial bond strength was calculated from Kelly's formula

$$\tau_{app} = C\frac{\sigma_l d}{2l},$$

where $\sigma_l$ is the strength of the fiber of the mean fiber length $l$, and $C$ is correction factor [15].

— *Microbond test*: specimens for this test were prepared on thin cover glass plates as described elsewhere [11, 14]. A mini tensile machine was used to produce tensile load. The free fiber end was glued to the metallic bar attached to the force sensor. The free fiber length was 0.2–2 mm. The testing rate was 0.25 μm/s to ensure slow crack propagation. The testing device was placed on the stage of an optical microscope equipped with a video camera connected to a computer, which

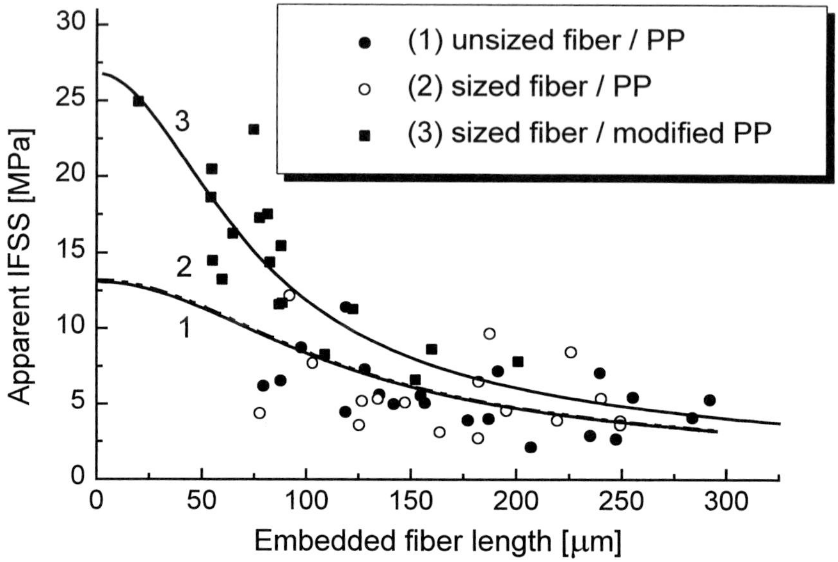


Fig. 1: Apparent interfacial shear strength in glass fiber–polypropylene systems plotted versus the embedded length: (1, 2) PP, (3) PPM. Results of the pull-out test; fiber sized with APS. Lines are approximations of experimental data according to shear lag analysis [21] using the least-squares method.

made possible to record the picture of crack propagation in step with the applied force. Thus, the crack length was obtained as a function of the applied external force, $F$. Three characteristic forces — the maximum force $F_{max}$, the debond force $F_d$, and the friction force $F_f$, — were determined for each trial. The debond force was obtained as a limit of the function $F(l_{cr})$, where $l_{cr}$ is the experimentally measured crack length [14]. The local adhesion strength, $\tau_{ult}$, the adhesional pressure, $\sigma_{ult}$, and the work of adhesion, $W_A$, were calculated according to the algorithm proposed in Ref. 14.

***Micro-thermal analysis***

Micro-thermal analysis μTA 2990 (TA Instruments) combines a scanning force microscope with the thermomechanical analysis. The experiments were made on the polished samples in the vicinity of the fiber-matrix interface. The glass transition temperatures were detected at the distances 2, 4 and 20 μm from the glass fiber surface.

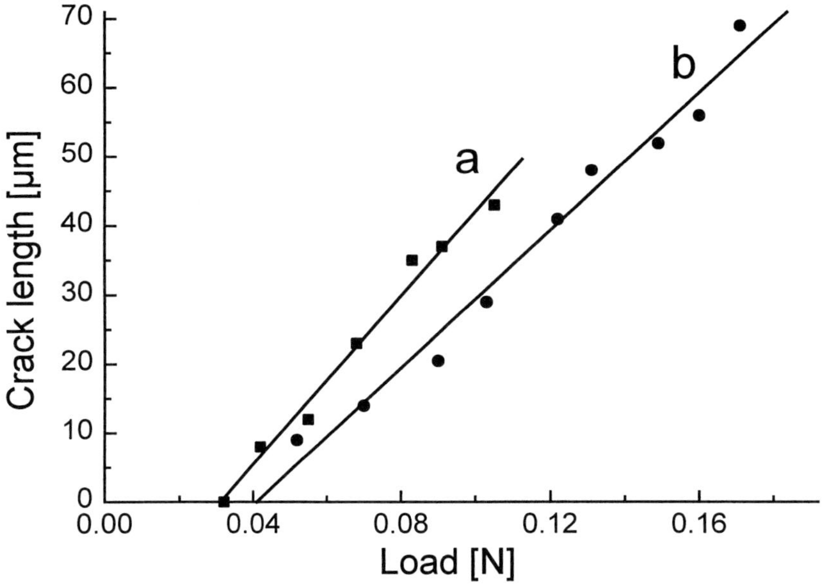


Fig. 2: Experimental relationships between applied load and the crack length from microbond test: (a) PP, (b)PPM; fiber sized with APS.

## Results and Discussion

### *Non-polar polymers*

The main problem of fiber reinforced non-polar polymers is low adhesion between the fiber and the matrix. Due to the lack of reactive groups in the polymer chains, the polymer adhesion to fiber surface is provided only by van der Waals forces that can hardly be influenced by means of interfacial design. Indeed, as was shown by pull-out [17)] and push-out [18)] experiments as well as by testing macrocomposites [19, 20)], any treatment of the fiber surface did not result in the improvement of the bond strength or composite properties in the case of PP or PE matrices.

Table 1. Effect of fiber coating on interfacial characteristics of the polypropylene/glass fiber system.

| Matrix | Fiber coating | $\tau_{ult}$ (MPa) | $\sigma_{ult}$ (MPa) | $W_A$ (mJ/m$^2$) |
|---|---|---|---|---|
| PP | — | 13.1 | 98 | 69 |
| PP | APS | 13.2 | 100 | 70 |
| PPM | — | 12.8 | 97 | 68 |
| PPM | APS | 26.8 | 147 | 91 |

The only possible way to improve adhesion of non-polar polymers is simultaneous modification of both fiber surface and the matrix in such a way that reactive acidic and basic groups are created on the contacting surfaces. This approach appeared very effective for PP composites: sizing of

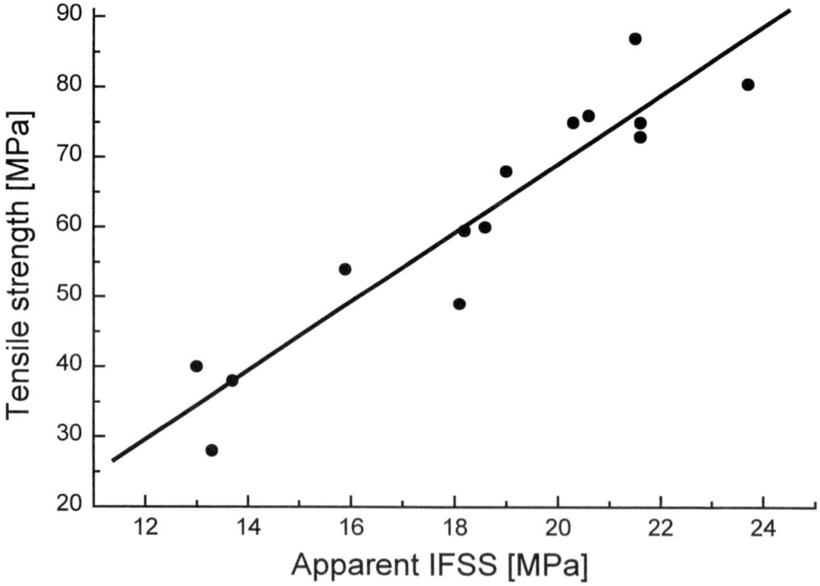


Fig. 3: Tensile strength of PP/glass fiber composites versus interfacial shear strength from pull-out test.

glass fibers by basic APS and simultaneous introducing acidic groups into the matrix resulted in considerable increase in the strength of composites with continuous [17] and short [2] glass fibers as well as in the bond strength from push-out test [18].

As can be seen from the results of pull-out test presented in Fig. 1, the interfacial shear strength is markedly greater for modified PP as compared to unmodified matrix in the whole range of embedded lengths. A crack between modified PP and sized fiber is initiated at considerably greater applied external force (Fig. 2). Interfacial characteristics for PP–glass fiber systems calculated from pull-out and microbond data are summarized in Table 1. It should be mentioned that the work of adhesion between PP and glass fiber surface estimated from micromechanical testing is in a good agreement with the value obtained from wetting and inverse gas chromatography [21]. An increase in the work of adhesion for modified PP–sized fiber system up to 91 $mJ/m^2$ can be attributed to the acid-base interactions between amino groups on the fiber surface and acidic groups in the polymer chains. A strong correlation between the strength of micro- and macrocomposites (Fig. 3) is an additional evidence for the fact that adhesional interaction is of key importance in the interfacial design in the case of non-polar polymers like PP.

The modification of the whole matrix material is not the only possibility to affect the adhesional interaction at the interface. Another way is introducing a layer of polar polymer between the fiber and the non-polar matrix. The interphase plays a key role in stress transfer, therefore a thin layer

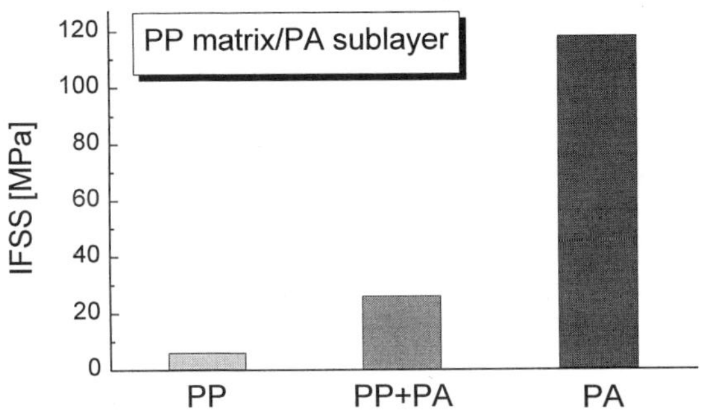


Fig. 4: Interfacial shear strength in carbon fiber / electrodeposited sublayer / polymer matrix systems. Results of the fragmentation test; left bar shows results for only matrix material, right bar, for only sublayer material.

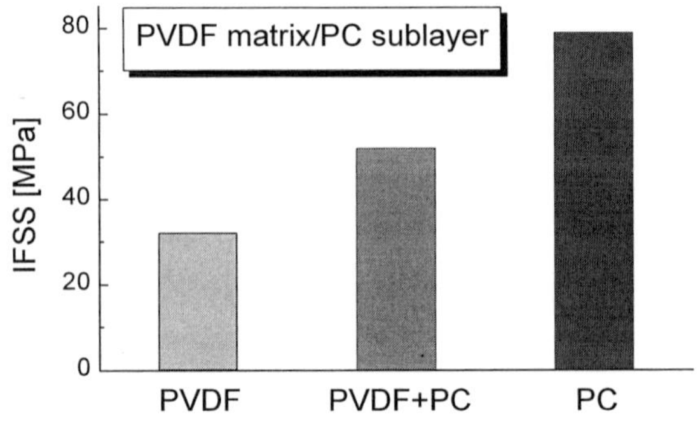


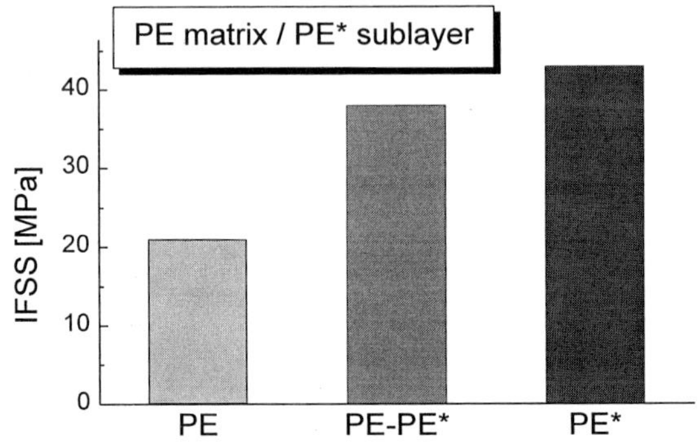


Fig. 4 (continued): Interfacial shear strength in carbon fiber / electrodeposited sublayer / polymer matrix systems. Results of the fragmentation test; left bar shows results for only matrix material, right bar, for only sublayer material. PE* is PE modified by gamma radiation in air.

of another polymer with good adhesional ability could be very effective for improving composite strength. In the case of continuous fibers such a layer can be applied by deposition of the powdered polymer onto fibers in an electric field [15]. This operation can be easily combined with the deposition of main matrix material used in dry technologies of composite fabrication.

When creating interphases using this technique, interdiffusion of the sublayer and the matrix polymer is facilitated by the fact that the process of interphase formation takes place for a sufficiently long time and at the temperature which is above melting points of both polymers. Bond strength measurement for various carbon fiber/sublayer/ matrix systems showed that in all cases it was intermediate between corresponding values for the matrix and sublayer materials

(Fig. 4). By using sublayers or interphases, two mechanisms of bonding should be taken into account: good adhesion to fiber surface is due to presence of active groups involved in acid-base interaction at the interface, while at the sublayer-matrix boundary interdiffusion plays the main part [20–22] .

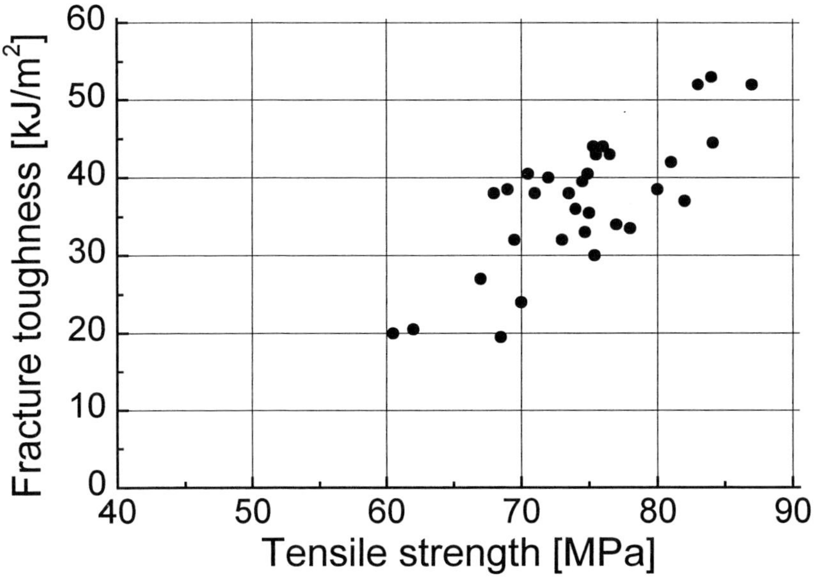


Fig. 5: Impact toughness versus tensile strength of PP/glass fiber composites with different fiber coatings.

The sublayers can also be applied from solutions or polymer dispersions; this procedure can be done immediately after the continuous fiber spinning process [16]. Having varied the chemical composition of film formers, we showed that it affected to a great extent the strength of PP-glass fiber composites [2, 17] . It should be mentioned that thickness of a sublayer and interdiffusion play a great part in such interphases [23]. When the sublayer is too thick, it becomes the "weak point" of the system, and the failure process occurs within it. Washing off the excessive sublayer material results in an increase of the bond strength [23]. According to recent investigation, the optimum interphase thickness for different polymer-fiber pairs is about 0.1–0.2 μm [4, 24]. Proper interfacial design should take into account not only acid-base interactions at the interface, but also thickness of the interphase and interdiffusion with bulk matrix material. Under these conditions, simultaneous increase in impact resistance and tensile strength of the composites can be achieved (Fig. 5).

To activate PE powder surface, maleic anhydride and styrene monomers were grafted onto it. The presence of oxygen-containing functional groups was determined using XPS. ToF-SIMS investigations were carried out to obtain information about the molecular build-up of the grafted polymers. The spectra gave evidence of polymer chains with alternating maleic anhydride and styrene units (Table 2).

Table 2. Some properties of modified polyethylene (PEMS) and its interface with glass fibers.

| Material | MAH (wt%) /ST (wt%)[a] | Gel content (%) | $\tau_{app}$ (MPa) | $\tau_f$ (MPa) | $\sigma_{ult}$ (MPa) | [C–O]/ [C=O][b] |
|---|---|---|---|---|---|---|
| PEMS1 | 5/0 | 0.5 | 4.5 | 2.7 | 53.6 | 0.497 |
| PEMS2 | 5/2.5 | 4.1 | 27.6 | 12.3 | 126.3 | 0.853 |
| PEMS2T | 5/2.5[c] | 11.8 | 30.0 | 20.0 | 152.8 | — |
| PEMS3 | 5/5 | 5.5 | 28.9 | 4.7 | 171.0 | 0.610 |
| PEMS3T | 5/5[c] | 11.0 | 31.0 | 17.5 | 135.8 | — |

[a] MAH — maleic anhydride; ST — styrene.
[b] From Ref. [25].
[c] Heat treatment for 10 min at 250°C.

The monomers grafted onto the PE surface increased its polarity. To investigate the effect of different chemical composition on the adhesion strength, single APS-sized glass fibers were embedded in their melts. The force-displacement curves showed significant changes of the interphase behavior (Fig. 6). First, the adhesion of all modified matrices was considerably greater than the adhesion of non-modified matrix. The subsequent thermal treatment caused, in all probability, cross-linking within the interphase and further increase in the number of local bonds at the interface. The increased degree of cross-linking was confirmed by greater gel content (Table 2). As a result, the interphase for thermally treated samples was much stronger, and the failure mechanism was quite different. As can be seen in Fig. 6, these samples failed through matrix yielding rather than brittle interfacial crack propagation.

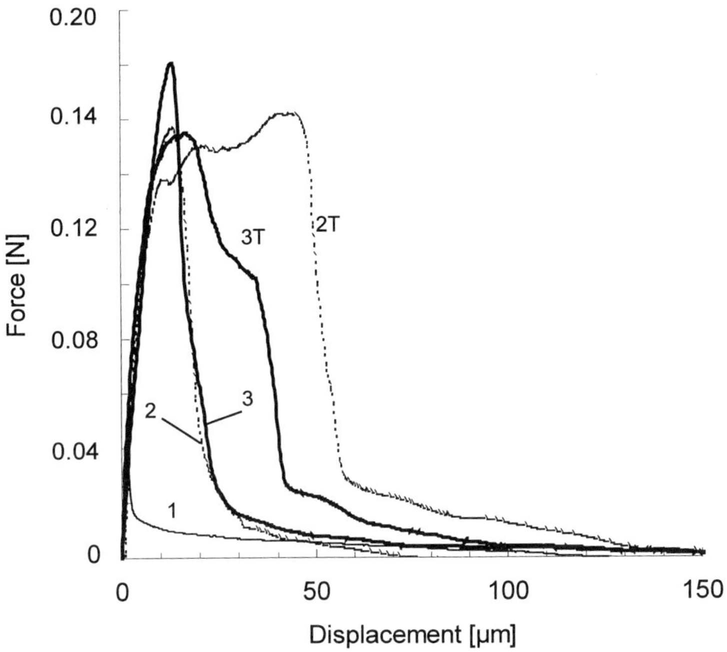


Fig. 6: Pull-out force-displacement curves for PEMS matrices / APS-treated glass fibers. Matrix designations (1, 2, 2T, 3, 3T) are the same as in Table 2.

***Polar thermoplastic polymers***

As was shown by recent research, acid-base interaction plays an important part in adhesion of polar polymers to reinforcing fibers; its contribution to the total bond strength and work of adhesion can be greater than the contribution of dispersion forces [21, 26)]. Thus, the creation of complementary donor or acceptor groups on the fiber surface should be a basis of interfacial design in this case. Indeed, it was demonstrated for many polar thermoplastic polymers that the bond strength of polymer-fiber joints was proportional to such thermodynamic parameters as the surface free energy or the reversible work of adhesion estimated from wetting [16, 27, 28)] or inverse gas chromatography [21, 26)].

Thus, a change of the work of adhesion after fiber treatment can be used to predict the trend for interfacial bond strength. For example, $W_A$ between glass fiber surface and ABS decreases after fiber sizing with APS (Table 3). The reason for this is a suppression of donor-acceptor interaction between acidic hydroxyl groups on the unsized glass surface and donor groups (nitrile, vinyl, phenyl) of the polymers. The sizing alters substantially the chemical nature of glass fibers: basic amino groups cover the acidic hydroxyl groups of the glass fiber surface [16, 17]. As a result, the

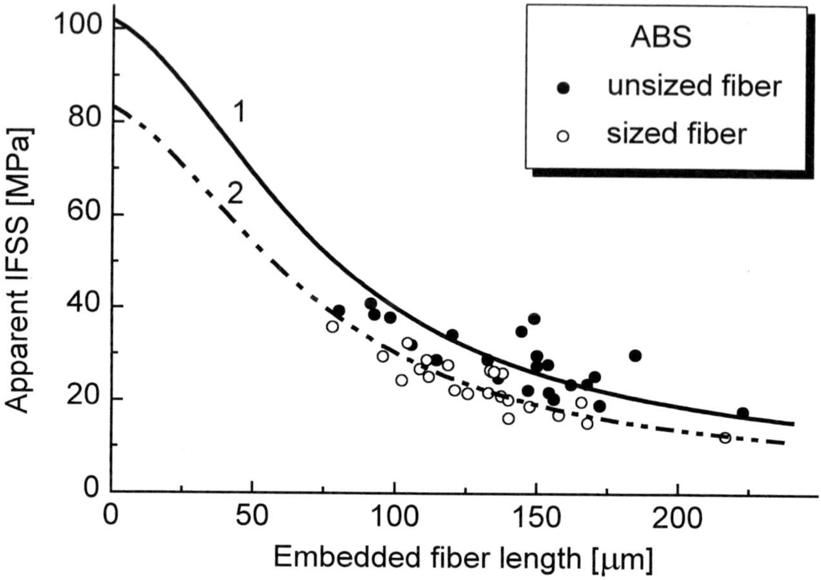


Fig. 7: Apparent interfacial shear strength in glass fiber–ABS systems plotted versus the embedded length: (1) unsized, (2) APS sized fiber. Results of the pull-out test. Lines are approximations of experimental data according to shear lag analysis [21] using the least-squares method.

interfacial bond strength also decreases upon fiber sizing (Fig. 7). The strength of macrocomposites made from ABS and aminosilane sized glass fibers was lower as compared to that of the composites with unsized fibers (Table 3, [29]).

Molecular weight of the polymer is also very important for interfacial design. In PS/glass fiber systems, the increase in average molecular weight led to enhanced adhesion strength. In contrast to previously reported decrease of adhesion strength when APS-sizing was used [21], already at moderate molecular weights an increase in adhesion strength was found (Table 4). This can hardly be explained by changes in polarity, however, could be due to creation of stronger interphases. Low molecular weight polymers usually fail through brittle cracks. In high molecular weight polymers, this mechanism is inhibited by interdiffusion of chains.

Table 3. Effect of the sizing on the work of adhesion, interfacial bond strength, and strength of short-fiber reinforced composites.

| Matrix | Fiber*a)* | $W_A$ (mJ/m$^2$)*b)* | $\tau_{app}$ (MPa) | Tensile strength (MPa) | Flexural strength (MPa) | Impact strength (kJ/m$^2$) |
|---|---|---|---|---|---|---|
| ABS | unsized | 102 | 31.8 | 68 | 101 | 6.8 |
| ABS | APS | 83 | 22.7 | 48 | 81 | 8.6 |
| PA | unsized | 83 | 24.6 | 101 | 172 | 7.1 |
| PA | APS | 73 | 33.6 | 128 | 224 | 11.2 |
| PA | APS+epoxy | 94 | 51.0 | 132 | 231 | 12.1 |

*a)* Fiber content: ABS, 15% [28]; PA, 30% [16, 29].
*b)* From IGC (ABS), from wetting data (PA) [16].

Table 4. Effect of fiber sizing and molecular weight distribution of the matrix polymer on adhesion strength between polystyrene and glass fibers.

| Matrix | | Sizing | $\tau_{ult}$ (MPa) | $\tau_{app}$ (MPa) |
|---|---|---|---|---|
| $M_n$ (g/mol) | $M_w$ (g/mol) | | | |
| 20,000 | 20,000 | — | 62.4 | 15.2 |
| 140,000 | 280,000 | — | 85.9 | 22.3 |
| 140,000 | 280,000 | APS | 91.0 | 23.5 |
| 1,800,000 | 1,800,000 | — | 96.5 | 26.1 |

A further effect of artificially created interphases could be studied by applying a sublayer of acid-terminated polystyrene (TEMPO-method after Fukuda et al. [30] ) of different molecular weight onto aminosilane sized glass fibers from 3 wt.% solutions of tetrahydrofurane (similarly to application of film formers during glass fiber spinning). Donor-acceptor interaction in this system occurs between COOH-groups of the sublayer and $NH_2$-groups from aminosilane. Thus, a precondition for covalent bonds formation was achieved, which resulted in a considerable increase in adhesion strength (Table 5). Adhesion strength appeared to be dependent on the molecular weight of the functionalized PS, and at the molecular weight of $M_n$ = 34,700 a clear maximum was detected. This effect can be due to the competition between interfacial adhesion and strength of the interphase. While the cohesive brittle fracture for smaller molecular weights has a greater influence of the adhesion strength than the adhesive acid-base interactions, the reduced adhesion strength with further enhanced molecular weights of the sublayer can be due to decreased number of acidic groups (end-terminated products).

Table 5. Effect of acid terminated PS sublayers on adhesion strength between polystyrene ($M_w$ = 280,000) and aminosilane sized glass fibers.

| Sublayer | | $\tau_{ult}$ (MPa) | $\tau_{app}$ (MPa) |
|---|---|---|---|
| $M_n$ (g/mol) | $M_w$ (g/mol) | | |
| — | — | 91.0 | 23.5 |
| 8,000 | 10,500 | 95.6 | 24.9 |
| 28,400 | 39,000 | 116.9 | 29.5 |
| 34,700 | 44,200 | 148.0 | 44.9 |
| 47,800 | 70,000 | 107.9 | 27.8 |

Thus, acid-base interactions are not the only mechanism for controlling interfacial adhesion and bond strength. It can be seen, for example, from the pull-out data obtained by testing joints of PA

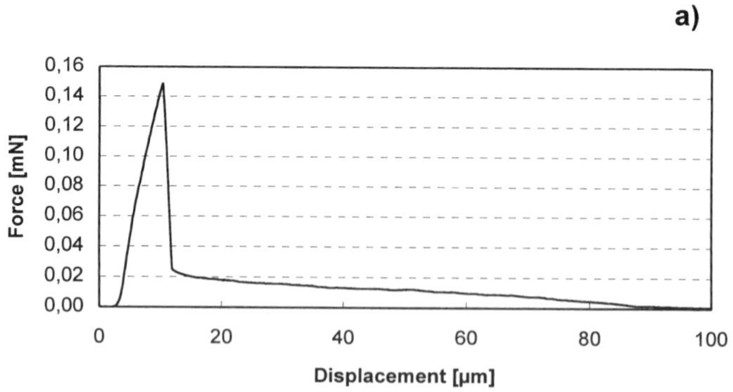


Fig. 8: Typical force-displacement curves from the pull-out test for glass fiber/polymer matrix systems: (a) ABS.

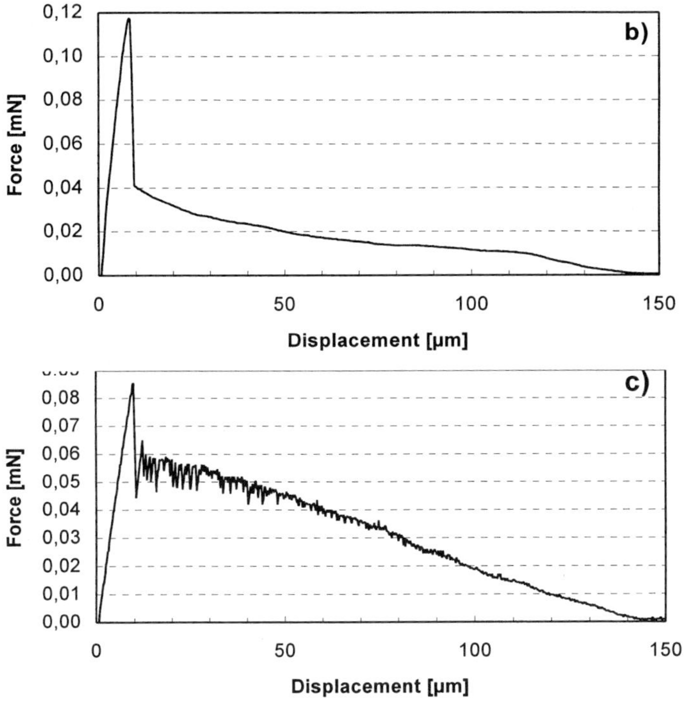


Fig. 8 (continued): Typical force-displacement curves from the pull-out test for glass fiber / polymer matrix systems: (b) PA, (c) PPS.

or PPS with glass fibers. Some reported data on the strength and adhesion in PA-glass fiber systems seem to be contradictory. So, on the one hand, an evident correlation has been found between $W_A$ from wetting data and composite strength for PA with unsized and sized glass fibers [16]. On the other hand, the use of some sizing agents resulted in a decrease of fiber surface energy or $W_A$, but at the same time the composite strength increased [20, 31]. A similar effect was observed for other thermoplastic polymers [22]. Our experiments have confirmed this discrepancy, which can be seen from Table 3. So, APS sized fiber has lower surface energy as compared to the unsized fiber, while the application of an epoxy-containing layer increases the surface energy drastically [31]. Nevertheless, both treatments improve tensile strength as well as impact resistance of short fiber reinforced PA composites.

The obvious difference in behavior of glass reinforced PA and ABS composites can be explained by the ability of PA of forming transcrystalline region around a fiber when preparing the composites [7, 32]. ABS is an amorphous polymer; in this case, the interfacial bond strength is governed by adhesion, especially by acid-base interactions. Microscopic observations showed that contact failure under loading occurred in adhesional way (through the interface), and post-debonding friction was very low (Fig. 8a). In contrast, transcrystalline layers of PA at glass fiber surface influence the stress transfer, giving rise to composite strength [32]. Adhesion of PA to glass surface is rather high [16, 31], therefore the failure within the transcrystalline interphase is probable. The high post-debonding friction confirms this assumption (Fig. 8b). Sizing glass fiber with APS as well as treatment by epoxy dispersion leads to increase in both interfacial strength and friction (Table 6). As a result, strength of macrocomposites is also enhanced (Table 3).

Table 6. Effect of the fiber sizing on the interfacial shear strength and frictional strength

| Matrix | Fiber | $\tau_{mean}$ (MPa) | $\tau_f$ (MPa) |
|---|---|---|---|
| PA | unsized | 24.6 | 12.2 |
| PA | APS | 33.6 | 18.5 |
| PA | APS+epoxy | 51.0 | 24.2 |
| PPS | unsized | 11.8 | 9.0 |
| PPS | APS | 17.5 | 12.7 |
| vinyl ester | MAS+UPE | 67.4 | 4.2 |
| vinyl ester | PVA | 22.5 | 9.5 |

Very high post-debonding friction was characteristic of PPS-glass fiber system (Fig. 8c). Sizing the fibers by APS resulted in slight enhancement of bond strength and interfacial friction (Table 6). This effect can not be explained by an increase of adhesion. Indeed, both ABS and PPS possess pronounced donor nature, thus their adhesional interaction with glass surface must decrease after aminosilane fiber sizing. This is the case for ABS, but not for PPS. The reason for this can be the same as for PA composites - namely, formation of a transcrystalline layer around the fibers. Recently, Meretz et al. [33] have shown that the bond strength between PPS and carbon fiber varied in a wide range depending on the thermal conditions of the contact formation. The failure mode changed from ductile to brittle indicating different structures of PPS layers adjacent to fiber surface. It should be mentioned that no fiber sizing was used in that work, so it can be regarded as an example of controlling "natural" interphases between a fiber and a thermoplastic polymer. Similar results were obtained for other thermoplastic polymer joints with carbon and glass fibers: the bond strength and the failure mode appeared to depend considerably on the conditions of adhesional contact formation [27, 34–37].

Table 7. Effect of the fiber sizing on the interfacial characteristics and strength of longitudinal glass fiber/epoxy composites. Fiber content is 50 %.

| Fiber | $\tau_{ult}$ (MPa) | $\sigma_{ult}$ (MPa) | Tensile strength (MPa) | Tensile modulus (GPa) | $\varepsilon$ (%) | Shear strength (MPa) |
|---|---|---|---|---|---|---|
| unsized | 59.7 | 153 | 27.3 | 8.7 | 0.3 | 55.2 |
| APS | 160.7 | 326 | 72.9 | 8.9 | 1.0 | 62.0 |
| APS+PU | 84.3 | 232 | 49.2 | 9.3 | 0.6 | 55.3 |
| PVAc | 51.0 | 185 | 24.1 | 8.2 | 0.3 | 54.7 |

Taking into account good interfacial adhesion of most polar thermoplastic polymers to the glass and carbon surfaces, we can define the aim of interfacial design for this case to be creation of sufficiently thick and strong boundary layers between the fiber and the matrix. Varying temperature and time during the contact formation can be a powerful tool of controlling structure and strength of interphases. In the case of designed interphases introduced between the fiber and the matrix, interdiffusion and post-debonding friction are very important. The friction after debonding plays a great part in the dissipation of the fracture energy preventing a composite from brittle fracture and providing high impact resistance [4].

***Thermosetting polymers***

Most thermosetting polymers possess very good adhesion to glass and carbon fiber surface, what results in the brittle fracture and low impact resistance of the composites [4]. The increase in interfacial bond strength upon fiber sizing is often followed by deterioration of the strength of macrocomposites [2, 4, 38]. Therefore, the objective of interfacial design in this particular case is different from that for thermoplastic polymers. Interphases are created in epoxy and other thermosetting polymers in a specific way: preferential adsorption on the fiber surface of one component of matrix composition [1, 3, 39, 40] and possible chemical bonding at the interface [1, 39–41] are very important. Control over these processes assumes knowledge of the structure and properties of the formed interphase. Drzal et al. [40] have demonstrated that preferential adsorption and higher diffusivity of amine as compared to epoxy monomer leads to creation of brittle interphases with high modulus at the carbon fiber surface. Recent direct measurements by AFM gave similar result also for glass [2] and copper [3] fibers: the modulus of the interphase was markedly higher than that of the bulk matrix.

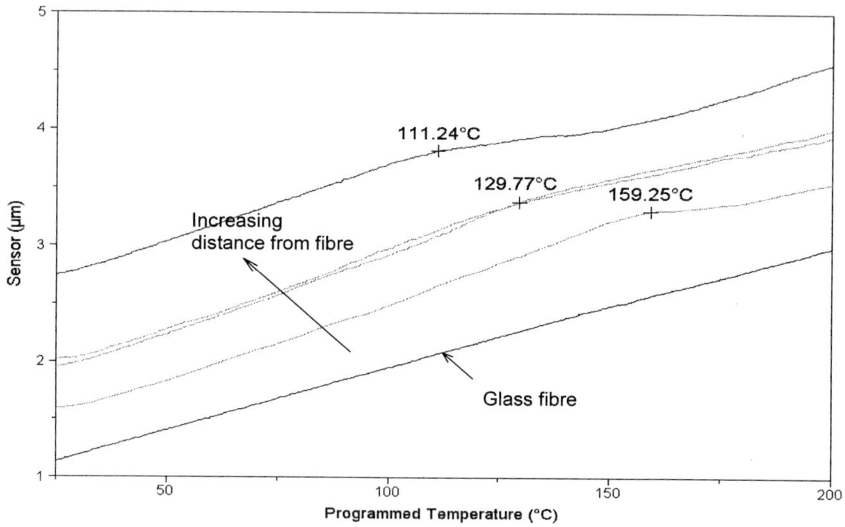


Fig. 9: μTA results for epoxy matrix at the following distances from glass fiber surface: (a) 2 μm, (b) 4 μm, (c) 20 μm.

Sizing of the glass fibers by aminosilane results in a considerable enhancement of interfacial strength [21, 42]. Transverse tensile strength and shear strength of long-fiber composites are also increased (Table 7). The possible reason for this can be covalent bonding between epoxy groups and amino groups on the fiber surface which can act as additional curing agent. Such a chemical interaction at the interface can contribute very much to bond strength: as was shown by Drzal et al. [41], a low (about 3%) chemical bonding increased interfacial strength between epoxy resin and carbon fiber by 34%.

Nevertheless, such an interphase remains very thin (APS does not cover the fiber surface) and brittle [2)]. The introduction of designed ductile and flexible interphases which could dissipate fracture energy seems to be more appropriate. With this aim, we used the combined APS + PU coating which could provide good adhesion and is rather ductile under loading. As was shown in our previous work [2)], the interphase formed between the epoxy matrix and the sized glass fiber was sufficiently stiff, thus providing good stress transfer through the interface. On the other hand, it appeared to be more uniform and ductile as compared to "natural" interphase for unsized glass fiber, with nearly linear gradient in nano stiffness and in friction coefficient [2)].

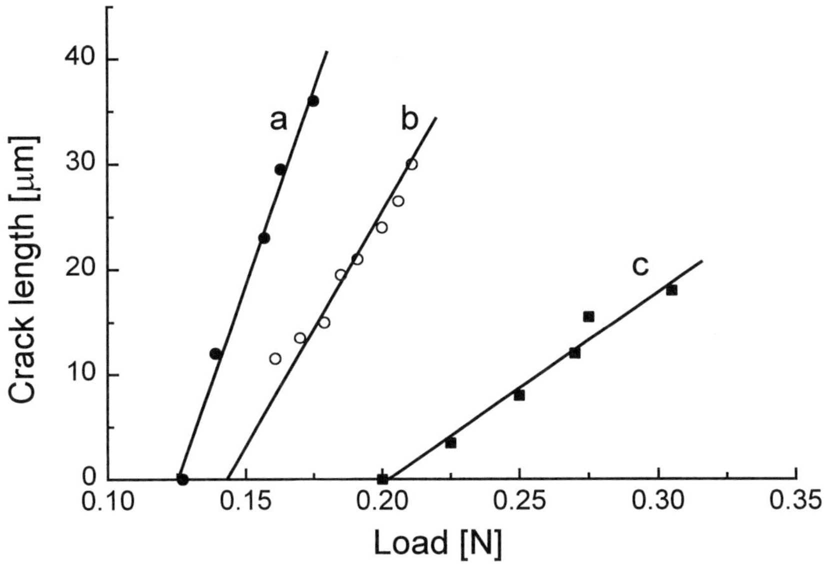


Fig. 10: Experimental relationships between applied load and the crack length from microbond test for glass fiber/epoxy matrix systems. (a) PVAc sizing, (b) unsized fiber, (c) fiber sized with APS.

The particular properties of this interphase can be confirmed by μTA results. Great local differences in the glass transition temperature were found between the interphase (at the distances 2μm and 4μm) from the fiber surface and bulk matrix (Fig. 9). The increase of $T_g$ was 48$^0$C which is in good agreement with the results obtained by Thomason [5)]. He has shown by DMA technique that $T_g$ in thin epoxy layers on the glass fiber is by 44$^0$C greater than in bulk matrix.

This effect was also described by Lipatov: $T_g$ of PS estimated from IGC data was by 30°C greater in thin layers on the glass surface than in the bulk [8]. This effect is due to interaction between the fiber surface and polymer macromolecules leading to a decrease of segmental mobility and thus to an increase in $T_g$. In the case of epoxy resin, an additional reason for this is enhanced cross-linking density at the fiber surface.

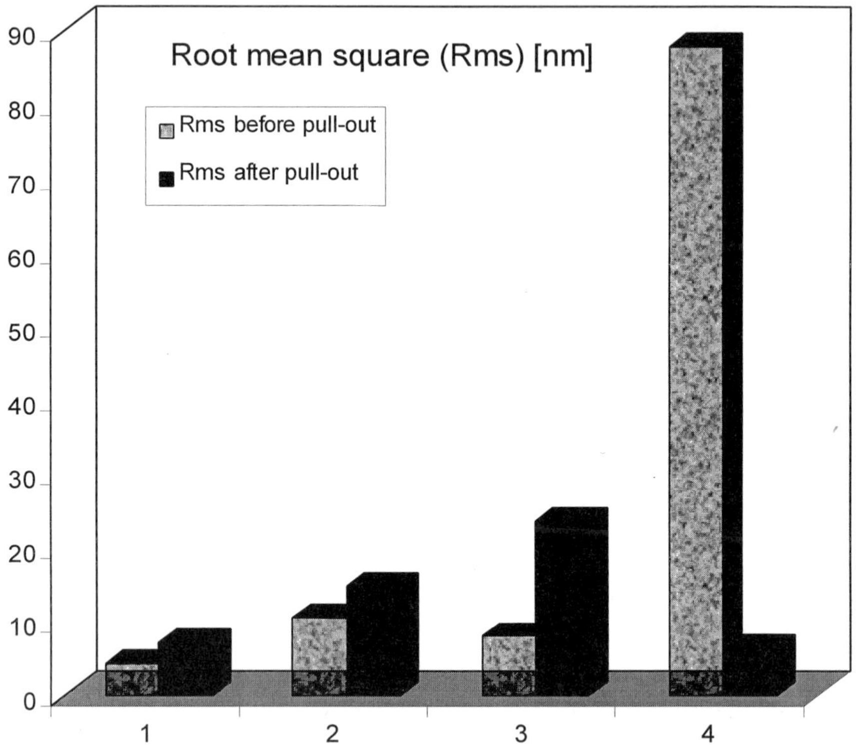


Fig. 11: Roughness parameters (nm) as root mean square ($R_{ms}$) $R_{ms} = \sqrt{\sum (Z_i - Z_{ave})^2 / N}$ of the glass fiber surface before and after the pull-out test: (1) unsized fiber, (2) APS, (3) APS+PU, (4) PVAc.

It is well known that fiber surface treatments affect the cure kinetics of anhydride- and amine-cured epoxy systems [43, 44]. A similar effect was recently described for the curing of vinyl ester resins on the surface of carbon [39] and glass [1] fibers. It was found that preferential adsorption of some constituents led also in this case to the creation of a "natural" interphase having properties different than the polymer bulk properties. This resulted in low mechanical properties of the composites due to low fiber-matrix adhesion. Application of fiber sizings was found to improve the level of adhesion between the carbon fiber and the matrix [39)].

The results of our pull-out experiments also demonstrated the strong dependence of the interfacial strength on the fiber sizing (Table 6). MAS sized glass fibers possess good adhesion to vinyl ester matrix, what can be explained by the ability of silane sizing to accelerate significantly the cure of vinyl ester [1)]. This cured layer at the fiber surface has evidently a high elastic modulus and thus provides an effective stress transfer. In contrast, application of PVA coating led to decrease in interfacial strength between vinyl ester and glass fiber (Table 6). It should be noted that the failure occurs in this case through the interface; the crack initiation can be seen by a clear kink on the force-displacement curve.

Also a PVAc coating seems to have low adhesion to glass surface; it can be confirmed by the results of pull-out and microbond tests for epoxy matrix (Table 7, [21)] ). The bond strength between PVAc sized glass fiber and the epoxy resin is low; the crack is initiated at smaller applied force (Fig. 10). Another possible reason for this can be weakness of the formed interphase itself - the ductile PVAc polymer cannot support a sufficiently high level of stress transfer.

The post-debonding friction is low for PVAc sized fibers. The application of APS sizing leads to an increase in the friction after pulling the fiber out from epoxy resin. It can be explained by the changed failure mechanism from pure interfacial debonding to plastic yielding of the matrix followed by cohesive failure within the interphase. This can be confirmed by the roughness parameters of glass fibers determined by AFM in tapping mode before and after pull-out test (Fig. 11). The roughness of APS and APS+PU sized fibers increased after pulling out from the epoxy matrix, what is evidence of a cohesive failure in the interphase. On the contrary, the roughness of the PVAc sized fibers was markedly greater before the test, and than it decreased to the level which was characteristic of unsized glass fibers. Therefore, the failure occurred through the fiber-coating interface.

Hence, in the case of thermosetting polymers particular mechanisms of the interphase formation, especially chemical reactions at the interface, should be taken into account. The introduction of "artificial" interphases is essential for preventing brittle failure. Such interphases should be sufficiently stiff, possess good adhesion to the fiber surface and provide the interdiffusion between matrix and coating.

## Conclusion

The possibilities of controlling bond strength between fiber surface and polymer matrix are discussed. The principles of interfacial design are different for the polymers having different chemical nature. Thus, the main issue in the case of non-polar polymers is an increase in the interfacial adhesion. This can be achieved by a chemical modification of the non-polar matrix, what provides acid-base interaction at the interface. Another possibility is the introduction of a reactive polymer sublayer between non-polar matrix and fiber surface.

For polar thermoplastic polymers the interfacial design is based on the controlling over acid-base interactions between fiber and matrix. The morphology of the "natural" interphases plays a great part in these systems. The varying temperature and time of contact formation can change drastically the structure of the boundary layer and affect the interfacial bond strength. The possible formation of a transcrystalline interphase should be taken into account.

In the case of thermoset polymers, preferential adsorption of one component of polymer composition and chemical reactions at the interface are of great importance. The "natural" interphases are often too brittle, therefore an introduction of ductile and flexible sublayers can be regarded as an important tool of the interfacial design. The thickness of the interphase and the possibility of interdiffusion at the matrix/sublayer interface should be taken into account by a fiber coating selection. An acceleration of the curing at the interface and possible chemical reactions between coating and matrix material result in improving interfacial bond strength.

## References


1. G. R. Palmese, O. A. Andersen, V. M. Karbhari, *Composites* **A30**, 11 (1999)
2. K. Mai, E. Mäder, M. Mühle, *Composites* **A29**, 1111 (1998)
3. M. R. VanLandingham, R. R. Dagastine, R. F. Eduljee, R. L. McCullough, J. W. Gillespie Jr, *Composites* **A30**, 75 (1999)
4. M. Labronici, H. Ishida, *Composite Interfaces* **2**, 199 (1994)
5. J. L. Thomason, *Composites Sci. Technol.* **44**, 87 (1992)
6. P. F. M. Meurs, P. J. G. Schreurs, T. Peijs, H. E. H. Meijer, *Composites* **A27**, 781 (1996)
7. R. L. Clark Jr, R. G. Kander, B. B. Sauer, *Composites* **A30**, 27 (1999)

8. Y. S. Lipatov, *Physical Chemistry of Filled Polymers,* engl. transl. by R. J. Mosely, Internat. Polym. Sci. and Techn. Monograph No 2 (1977), Moscow
9. V. I. Irzak, L. I. Kuzub, *Composite Interfaces* **4**, 45 (1996)
10. L. E. Asp, L. A. Berglund, R. Talreja, *Composites Sci. Technol.* **56**, 657 (1996)
11. A. Hampe, C. Marotzke, *J. Reinf. Plastics and Compos.* **16**, 341 (1997)
12. M. R. Piggott, Y. Xiong, *Composites Sci. Technol.* **52**, 535 (1994)
13. M. C. Andrews, R. J. Young, J. Mahy, *Composite Interfaces* **2**, 433 (1994)
14. S. Zhandarov, E. Pisanova, K. Schneider, *J. Adhesion Sci. Technol.,* **14**, 381 (2000)
15. S. Zhandarov, V. Dovgyalo, E. Pisanova, *J. Adhesion Sci. Technol.,* **8**, 995 (1994)
16. E. Mäder, K. Grundke, H.-J. Jacobasch, G. Wachinger, *Composites* **25**, 739 (1994)
17. E. Mäder, H.-J. Jacobasch, K. Grundke, T. Gietzelt, *Composites* **27A**, 907 (1996)
18. G. Kalinka, A. Leistner, A. Hampe, *Composites Sci. Technol.* **57**, 845 (1997)
19. J. M. Felix, P. Gatenholm, H. P. Schreiber, *Polymer Composites* **14**, 449 (1993)
20. I. Ulkem, H. P. Schreiber, *Composite Interfaces* **2**, 253 (1994)
21. E. Pisanova, E. Mäder, *J. Adhesion Sci. Technol.,* **14**, 415 (2000)
22. P. H. Harding, J. C. Berg, *J. Adhesion Sci. Technol.,* **11**, 471 (1997)
23. E. Mäder, E. Pisanova, *Polymer Composites,* in press
24. E. Pisanova, E. Mäder, S. Zhandarov, *ICCM-12: Proc. 12 Int. Conf. Composite Interfaces,* paper 419, ISBN 2-9514526-2-4, Paris (1999)
25. S. Schneider, PhD thesis, Technical University Dresden, Germany, 1997.
26. M. Nardin, J. Schultz, *Composite Interfaces* **1**, 177 (1993)
27. P. Bourgeois, T. Davidson, *J. Adhesion* **45**, 73 (1994)
28. Y. L. Zou, A. N. Netravali, *J. Adhesion Sci. Technol.,* **9**, 1505 (1995)
29. S. Y. Fu, B. Lauke, *J. Mater. Sci. Technol.* **13**, 389 (1997)
30. H. Malz, PhD thesis, Technical University Dresden, Germany,1999.
31. H.-J. Jacobasch, K. Grundke, E. Mäder, K.-H. Freitag, U. Panzer, *J. Adhesion Sci. Technol.,* **6**, 1381 (1992)
32. R. L. Clark Jr, M. D. Craven, R. G. Kander, *Composites* **30A**, 37 (1999)
33. S. Meretz, W. Auersch, C. Marotzke, E. Schulz, A. Hampe, *Composites Sci. Technol.* **48**, 285 (1993)
34. E. Pisanova, S. Zhandarov, V. Dovgyalo, *Polymer Composites,* **15**, 147 (1994)

35. M. C. Waterbury, L. T. Drzal, in: *Controlled Interfaces in Composite Materials,* H. Ishida (Ed.), Elsevier Science, New York 1990, p. 731

36. M. C. Paiva, M. Nardin, C. A. Bernardo, J. Schultz, *Composites Sci. Technol.* **57**, 839 (1997)

37. J. Denault, *Composite Interfaces* **2**, 275 (1994)

38. L. Ye, A. Afgani-Khatibi, G. Lawcock, Y.-W. Mai, *Composites* **29A**, 1525 (1998)

39. M. J. Rich, S. Corbin, L. T. Drzal, *ICCM-12: Proc. 12 Int. Conf. Composite Interfaces,* paper 772, ISBN 2-9514526-2-4, Paris (1999)

40. L. T. Drzal, M. Rich, M. Koenig, P. Lloyd, *J. Adhesion,* **16**, 133 (1983)

41. L. T. Drzal, N. Sigura, D. Hook, *Composite Interfaces* **4**, 337 (1997)

42. F. M. Zhao, N. Takeda, K. Inagaki, N. Ikuta, *Advanced Composites Letters,* **5**, 113 (1996)

43. G. R. Palmese, R. L. McCullough, *J. Adhesion,* **44**, 29 (1994)

44. A. Garton, W. T. K. Stewenson, *J. Polym. Sci. Polym. Chem.,* **26**, 541 (1988)

# Modification of polymers in the melt

Gudrun Schmidt-Naake*, Hinnerk Gordon Becker, Marcus Klak

Institut für Technische Chemie, Technische Universität Clausthal
Erzstraße 18,D-38678 Clausthal-Zellerfeld,Germany

SUMMARY: The kinetics of polymer-analogue reactions in the melt are discussed with the example of conversions of styrene copolymers including anhydride or nitrile groups in the side chain and additives with amino and hydroxyl groups. Poly(styrene-co-maleic anhydride) has been modified with primary alcohols and amines, secondary amines, diamines, amino acids and amino alcohols. Poly(styrene-co-acrylonitrile) has been modified with various amino alcohols.

## Introduction

To widen their application window, chemical modifications of the industrial standard polymers are necessary. Polymer modification in the melt can be performed through batch-mixing or extrusion (continuos). The reactive extrusion combines the chemical reaction like, for example, polymerisation or polymer modification (polymer-analogue reactions) with mechanical processes like melting, formation of polymer blends, recreation of hypermolecular structures and the forming process. By way of contrast to chemical reactions in solution or suspension, reactive extrusion needs no solvents or detergents and even no precipitation or removal of the solvent. The reaction velocity is very much higher than in solution due to the high reactant concentration and the higher temperatures in polymer melts. Otherwise a chemical process which should be performed by reactive extrusion or melt-batch-mixing has to achieve special requirements like miscibility of all reaction components, the ability to achieve a homogenous mixture and the control over thermal degradation. An important point is the analysis and control of the molecular weight to avoid serious polymer degradation. Furthermore the transformation of a reaction, well known in solution, into molten state has to take into account, that the processing parameters, which have a high influence on the reaction have to be adapted. Therefore the analysis of the kinetic course of the reaction under the new process conditions is necessary.In the last 15 years the reactive extrusion has obtained more and more industrial acceptance. There are more than 100 patent applications since 1980. The aim of our work [1-6] is the investigation of polymer analogue reactions between styrene

 CCC 1022-1360/00/$ 17.50+.50/0

copolymers with anhydride or nitrile groups in the polymer side chain and additives with hydroxyl- and amino-groups.

## Materials:

### a)Conversions with poly(styrene-co-maleic anhydride):

We used four different compositions of poly(styrene-co-maleic anhydride) (S/MA) with 16 wt% MA (S/MA-16), 24wt% MA (S/MA-24), 32 wt% MA (S/MA-32) and with 49wt% MA (S/MA-49). The S/MA-16 showed a glass transition temperature (Tg) of 138 °C, the S/MA-24 of 144°C, the S/MA-32 of 152 °C and the S/MA-49 of 160°C. The polymers were purchased as industrial products and used without further purification. Even the used additives (see Table 1) were not further purified.

### b) Conversions with poly(styrene-co-acrylonitrile):

The poly(styrene-co-acrylonitrile) (S/AN) contained 23,8 wt% (38 mole%) acrylonitrile (AN) and had a weight average molecular weight of 102.000 g/mole and a number average molecular weight of 39200 g/mole (polydispersity=2,18) and shows a Tg of 112 °C. The S/AN and all used additives (see Table 1) were used without further purification.

| | S/MSA | | | | | |
|---|---|---|---|---|---|---|
| **coholes** | $HO-[CH_2]_{11}-CH_3$ dodecyl alcohol | 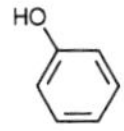 phenol | 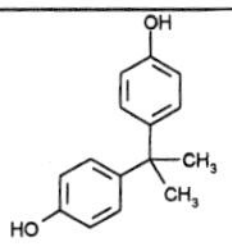 bisphenol A | | | |
| **mino- coholes** | 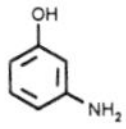 3-amino-phenol | 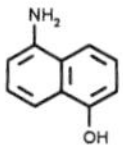 5-amino-naphthol | $H_2N-[H_2C]_3-OH$ 3-amino propanol | $HO-[CH_2]_2-NH_2$ 2-amino-ethanol | 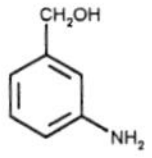 3-amino-benzyl-alcohol | |
| **prim. mines** | 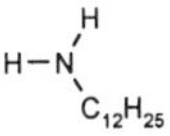 1-amino-dodecane | 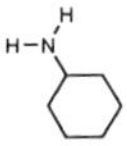 cyclohexyl-amine | 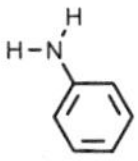 aniline | 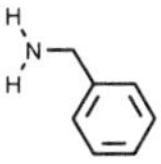 benzylamine | 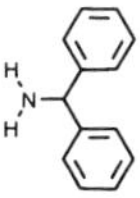 amino-diphenyl-methane | 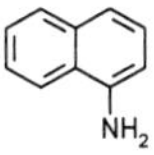 naphthyl-amine |
| **sec. mines** | 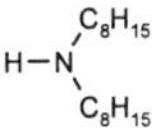 dioctyl-amine | 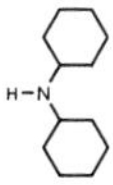 dicyclo-hexylamine | 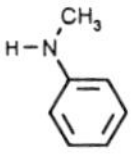 N-methyl-aniline | 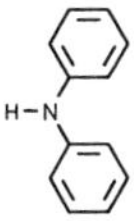 diphenyl-amine | | |
| **mino acids** | $H_2N-CH_2-COOH$ amino-acetic acid | $H_2N-[CH_2]_2-COOH$ 3-amino-propanoic acid | $H_2N-[CH_2]_5-COOH$ 6-amino-hexanoic acid | $H_2N-[CH_2]_{10}-COOH$ 11-amino-undecanoic acid | 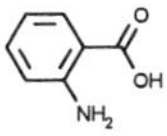 2-amino-benzoic acid | 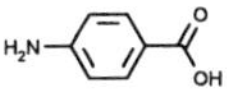 4-amino-benzoic acid |
| **mino iacids** | 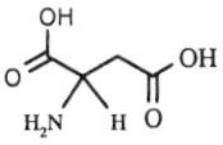 D, L-aspartic acid | 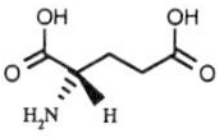 L-glutamic acid | 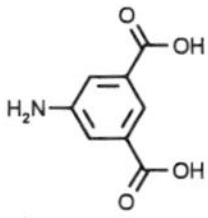 5-amino-isophthalic acid | 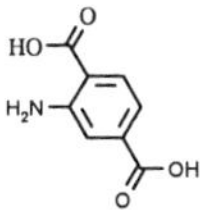 2-amino-terephthalic acid | | |

| SAN | | | | | | |
|---|---|---|---|---|---|---|
| **amino-alcohols** | 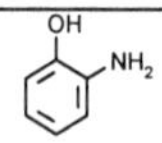<br>2-amino phenol | $HOH_2C-C(C_2H_5)(NH_2)-CH_2OH$<br>2-amino-2-ethylpropane -1,3diol | $HO-[CH_2]_2-NH_2$<br>2-amino-ethanol | 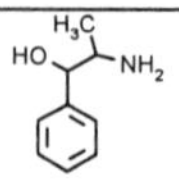<br>2-amino-1-phenyl-1propanol | $C_2H_5-CH(NH_2)-CH_2OH$<br>2-amino-1-butanol | $H_2N-[H_2C]_3-OH$<br>3-amino-propanol |

**Table 1**: Additives used for modification

## Reaction Procedure

For the chemical modification of all polymers we used a BRABENDER W30-EHT batch mixer with two sigma-rotors having a capacity of 30 cm³. The initial mixing temperature was 190 °C (S/AN) respectively 200 °C (S/MA), the rotation speed was 35 r.p.m.

The mixing (including premixing) procedure of polymer, additive and catalyst was optimized respectively for every polymer-additive group, resulting in different treatment. An example with a torque-time-diagram for solid additives is shown in Figure 1.

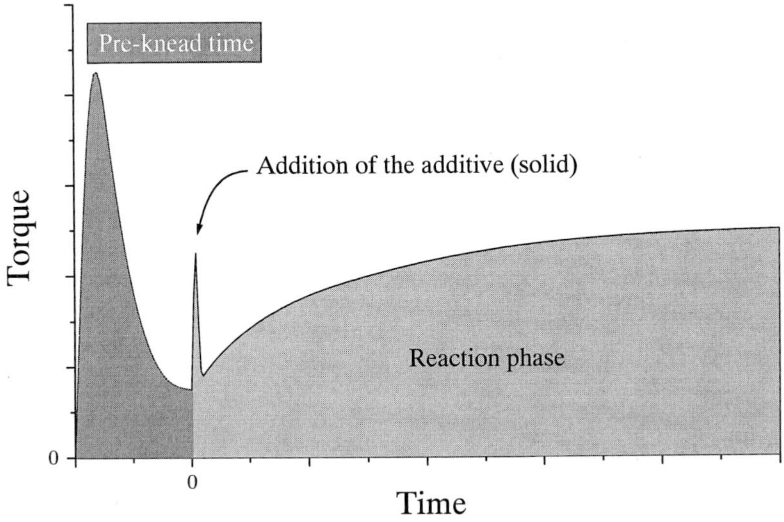


**Figure 1**: Torque -time-diagram for the addition of solid additives.

We tried to fill the reactor completely in order to get a better mixing. In most cases we used 20 g polymer and varied the amounts of additive and catalyst.

## Analysis of the reaction products

The purification of the samples was done by dissolving them in acetone followed by precipitation in methanol or water.

### a) Conversions with poly(styrene-co-maleic anhydride):

To determine the solubility, 1g of the product was kept in 100 ml acetone and stirred well for 24 h at room temperature. The sample then was centrifuged for 20 minutes (4500 rpm). The residue was dried under vacuum until its weight was constant and the soluble part was precipitated as described above. The conversion was only determined for the soluble part of the samples. To determine the conversion we used infrared (IR) spectroscopy (Fourier Transform Infrared Spectrometer NICOLET 5DXB) and nitrogen elemental analysis (method of Kjeldahl and VARIO EL from ELEMENTAR). Typical product-IR-peaks were at 1710 $cm^{-1}$ (imid) and 1950 $cm^{-1}$ (ester). For our rheological measures we used a Dynamic Analyser

RDA II from RHEOMETRICS. Our DSC data were generated on a DSC 820/25 from METTLER TOLEDO.

**b) Conversions with poly(styrene-co-acrylonitrile):**

The nitrile conversion was qualitatively determined by using IR spectroscopy (see above). Therefore the product peaks of the oxazoline at 1660 $cm^{-1}$ (-C=N-) and 1220 $cm^{-1}$ (C-O-C) were used as indicators. Additional specific additive peaks (i.e. OH- for 2-amino-2-ethylpropane-1,3diol (AEPD)) can be equally used. The quantitative determination of the oxygen content was done by elemental analysis (VARIO EL from ELEMENTAR).

## Results and Discussions

### 1. Conversions of S/MA with primary alcohols

In [1] we have already discussed the results of conversions between S/MA with various MA-content (16 and 32wt%) and dodecanol in molten state. The reaction of anhydrides with alcohols is reversible. Scheme 1 shows the reaction mechanism.

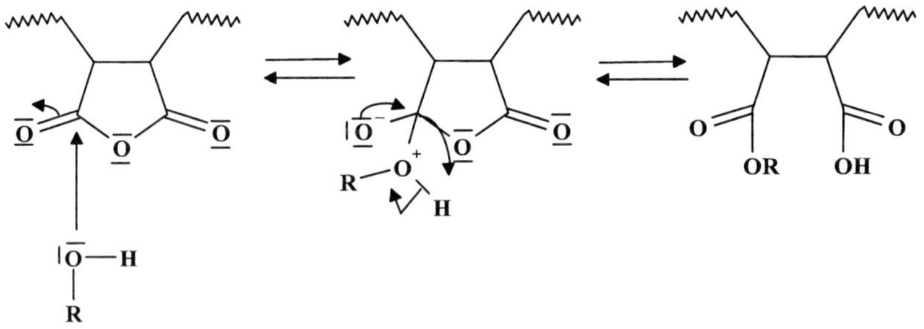

**Scheme 1**: Alcoholysis of anhydride groups in polymers.

If the reaction is performed in solution there has to be an excess of alcohol and catalyst to achieve a full conversion of the carbonyl groups. Therefore reaction times of more than 100 hours are necessary. In molten state only one of the two anhydride-carbonyl groups reacts. While the reachable conversion limit is not influenced by the anhydride content of the polymer, the reaction velocity increases with an increasing MA-content of the polymer, because of the higher concentration of reactive groups and of the autocatalytic effect of the constituted acid groups. The reaction can be catalysed with dibutyltinn oxide (DBTO) and itaconic acid (ITA) [1].

## 2. Conversion of S/MA with secondary amines

The reaction of S/MA-copolymers and secondary amines (scheme 2) shows a conversion limit after 10 minutes of reaction time. The limit is due to the steric hindrance of the used additives.

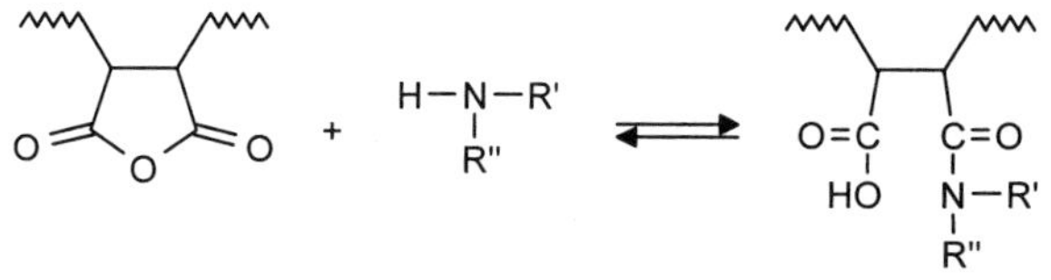


**Scheme 2**: Reaction of S/MA with secondary amines.

The reachable conversion limit in Figure 2 shows the influence of increasing anhydride-concentration in the used S/MA-polymers. In Figure 3, the influence of the batch ratio of additive groups and MA-groups is shown. An increase of the temperature (+ 20 °C) does not result in a higher conversion limit. The MFI increases with increasing rate of conversion.

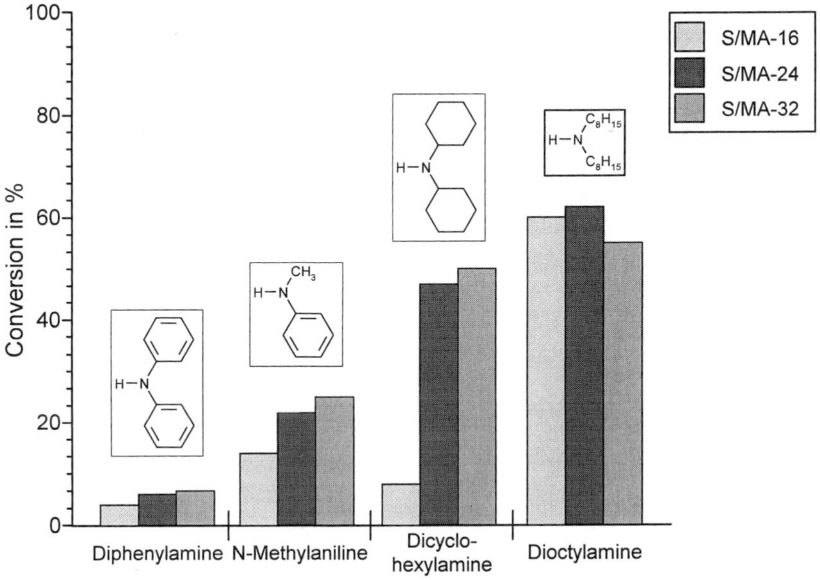


**Figure 2**: Reaction of S/MA with secondary amines ,$\vartheta$ = 200 °C, [MA]:[amine] = 1:1

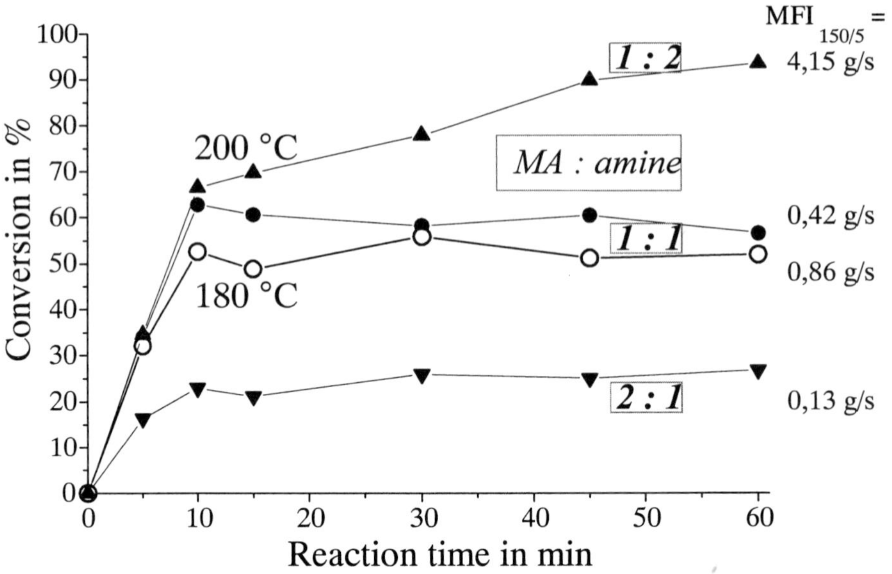


**Figure 3**: Conversion-time behaviour during the conversion of S/MA-16 with dioctylamine under variation of the MA:amine-ratio and at different temperatures.

## 3. Conversion of S/MA with primary amines and amino acids

The reaction of S/MA-polymers with primary alcylamines [8)] and ammonia [9-14)] in the melt, using batch-mixer and corotating extrusion has already been published. The imidisation of the anhydride group of S/MA runs in a heterocyclic manner resulting in cyclic imids after 5 minutes at 200 °C. The conversion with amino acids is done in an analogous way.

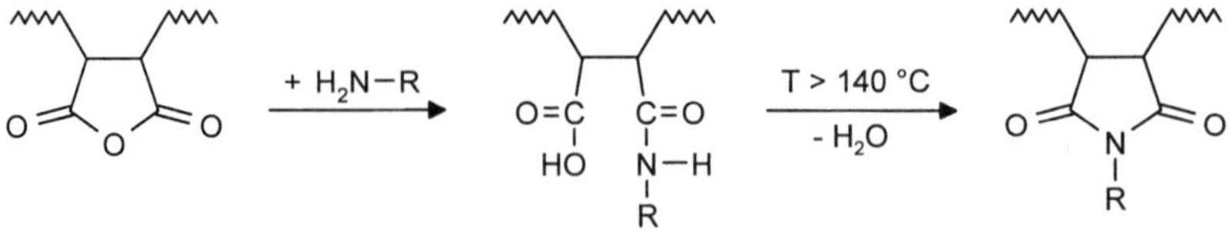


**Scheme 3**: Conversion of S/MA with primary amines and amino acids

It is noteworthy that there is an increase of the conversion limit with primary amines compared with secondary amines if S/MA-24 is used instead of S/MA-16, but a decrease with higher MA contents due to the increase of steric hindrance of the MA-groups. The conversion limit of reactions with primary amines decreases in the line 1-amino dodecane>amino diphenylmethane>benzyl amine>aniline>cycohexyl amine>naphthylamine (see Figure4). This can be explained with the higher steric room solidity.

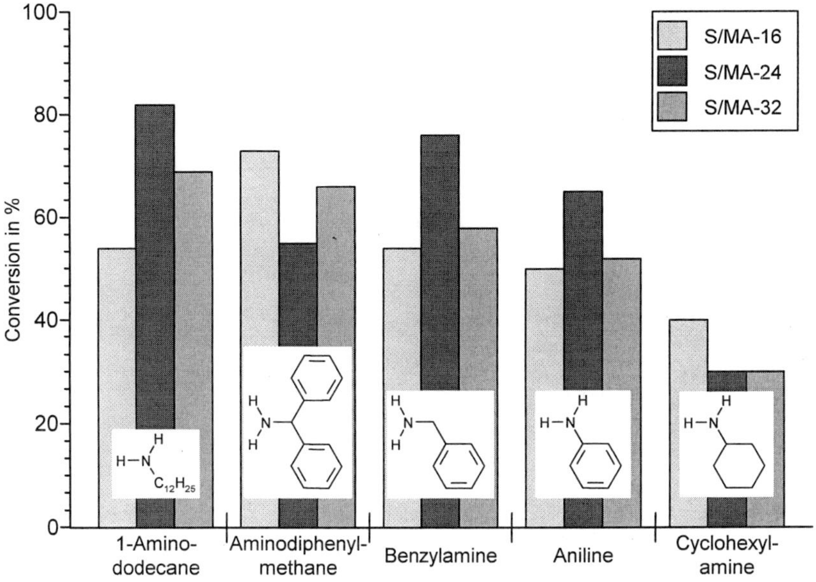


**Figure 4:** Conversion of S/MA with primary amines, $\vartheta$=200 °C, MA:amine=1:1.

The conversion limit decreases with an increase of additive concentration. The influence of temperature is small. Polymers with imid structures on side chains show an increase of the thermal shape stability, a good adhesion on metal surfaces and good electro-isolating properties. The glass transition temperature which can be used as expression for the thermal shape stability is shown in Figure 5 with variation of the imid content and imid structure.

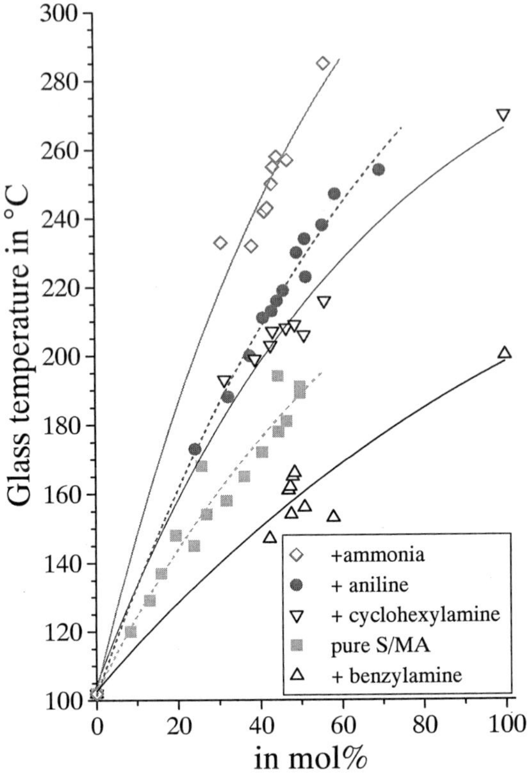


**Figure 5**: Coherence of glass transition temperature (Tg), imid structure and imid content of styrene-maleicimide copolymers.

The conversion of S/MA in solution [10-12] and in molten state results in comparable products [13]. In Figure 6 the similarity of reactions with S/MA and aniline in solution and in the molten state is shown on the example of Tg-increase during the reaction.

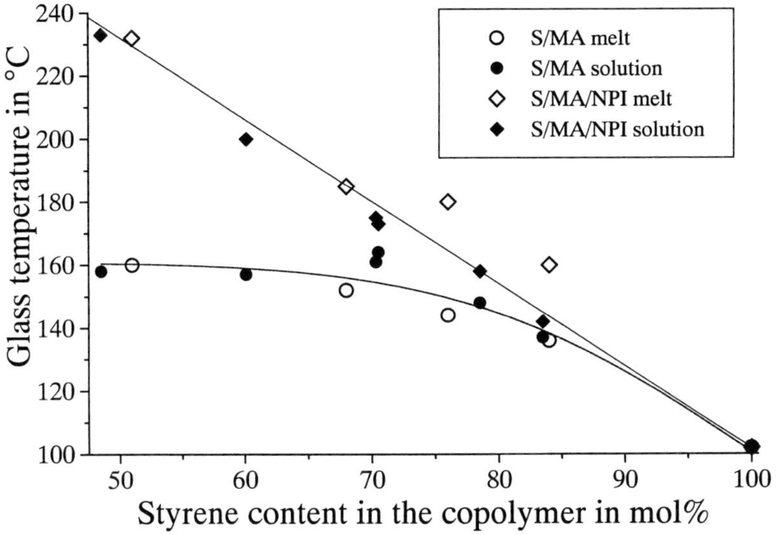


Figure 6: Glass temperatures of the reaction products from the conversion of S/MA copolymers and aniline in melt and solution.

Reactions of S/MA with amino acids result in products with aliphatic or aromatic carbonyl-groups in the side chain of the polymer [4,5]. These polymers can be transferred into water soluble products (polyelectrolytes) by the use of NaOH. Another noteworthy possibility is the transformation into acid chlorides and the use for reactive blending i.e. with polyamides. The chain length of the aliphatic amino acids does not influence the conversion limit (about 80 %). An exception is the reaction behaviour of amino acetic acid which transforms into 2,5-diketo piperazine if heated (Scheme 4).

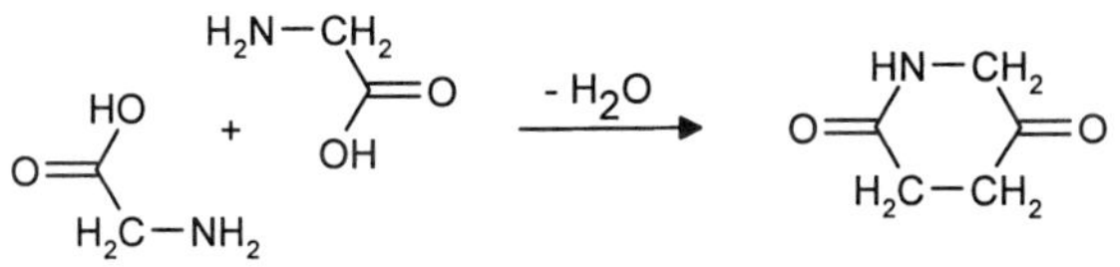


**Scheme 4**: Side reaction of amino acetic acid.

Vermeesch [8] reported a linear coherence between the chain length of the allylic side chains, that means the chain length of the used amines, and the Tg in reactions of aliphatic amines with the terpolymer MA/MMA/S (25 mol% MA, 7mol% MMA, 68 mol% S) in the melt. This can be explained with an increase of the free volume simultaneous with reduced packing density and decreased steric hindrance of the segmental mobility.

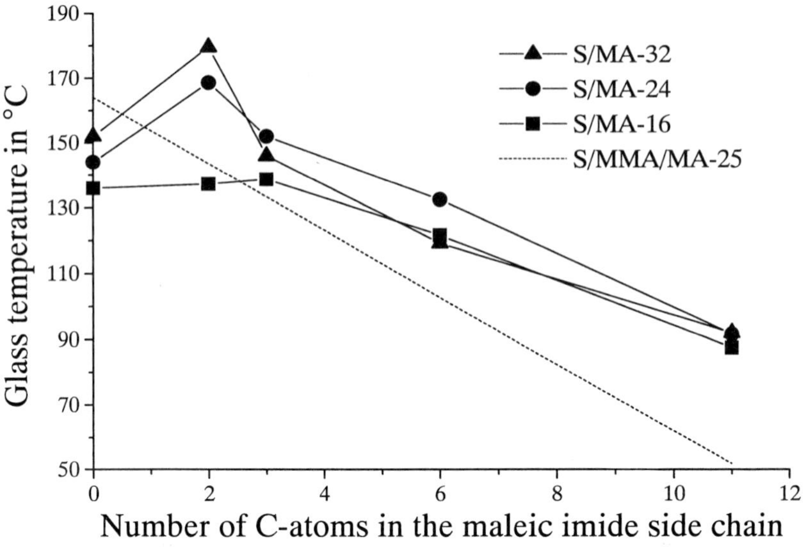


**Figure 7**: Glass transition temperatures (Tg) of the reaction products compared with the number of carbon atoms in the ω-amino acids used.

Figure 7 shows that the reaction of S/MA and ω-amino acids with various chain length results in similar products. An exceptance is again amino acetic acid (see above). The aromatic amino acids show a similar reaction behaviour with conversion limits of about 70 % and glass transition temperatures which are placed about 20 °C higher compared with the analogue non-substituted imides [4,5].

## 4. Conversion of S/MA with diamines

Polymer modification of S/MA in the melt using diamines results (even with low additive concentrations of MA: additive = 100:1) in highly crosslinked products (see Scheme 5). The melt transforms into solid powder in a very short time. The crosslinked products can be sulfonated and used as ion exchange media in water conditioning processes.

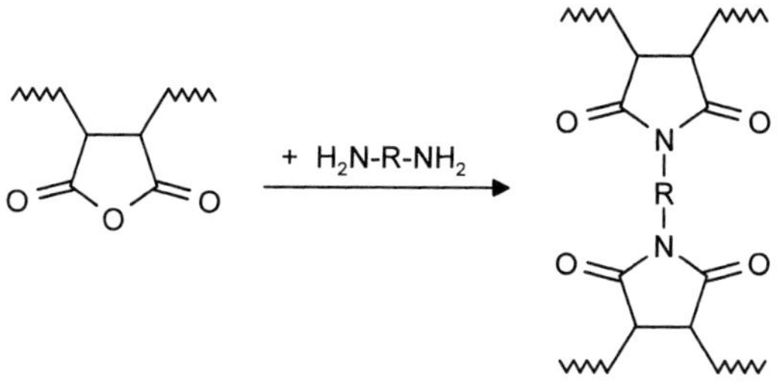


**Scheme 5**: Reaction of S/MA with diamines.

## 5. Conversion of S/MA with amino alcohols

The reaction of S/MA with amino alcohols should result in a fast imidisation which is followed by a slower alcoholysis (see Scheme 6). Since the alcoholysis results in a intra- and intermolecular crosslinking the degree of crosslinking should increase with the reaction time, that means the crosslinking should be controllable and lead to products with various crosslinking density in which the crosslinking can be completed in line with the forming process.

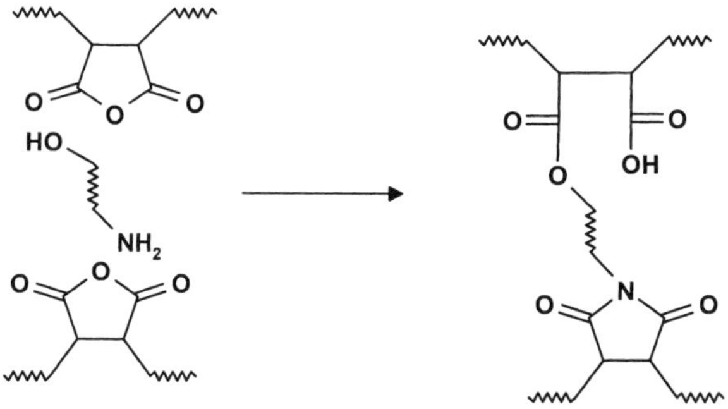

**Scheme 6**: Reaction of S/MA with amino alcohols.

The rheological analysis of the crosslinked products (see Table 2) shows a strong increase of the complex viscosity and the storing module G' i.e. an increase of the steadiness compared to the pure S/MA. Simultaneously there is a strong decrease of the loss factor tan δ i.e. a decrease of the viscous material properties. These results are a further indication of the increasing degree of crosslinking. A completely crosslinked product should show a solid-sphere viscosity (> $10^{-7}$ P) and a tan δ of nearly 0.

| | S/MA (84/16) | S/MA(84/16) + 3-amino-propanol ($MA:NH_2$=1:0,5) | Δ |
|---|---|---|---|
| **Complex Viscosity $\eta^*$ (T= 120 °C) [P]** | **216200** | **518800** | **+ 140 %** |
| **Loss Factor tan δ (T=120 °C)** | **0,99** | **0,21** | **- 80 %** |
| **Storing Module G` (T=120 °C) [dyn/cm²]** | **966200** | **3185800** | **+230 %** |

**Table 2**: Rheologic Data from the reaction of S/MA(84/16) with 3-amino-propanol (APr).

With the aromatic amino alcohols 3-amino phenol (3AP) and 5-amino naphthol (5AN) we got different results under similar reaction conditions. Amino phenol is the simplest aromatic amino alcohol. The meta-form (3AP) was chosen to avoid the steric hindrance of the ortho-form and the chinoid structures of the para-form of amino phenol. The conversion of S/MA-16 with 3AP results in completely soluble products. Even the infrared spectroscopy shows only the imid peak at 1710 cm-1 as product peak so that we suggest that only the aminolysis has taken place. The conversion limit was reached after a few minutes and amounted to 60 % ($MA:NH_2$= 1:1; maximum conversion=100 %) and 30 % ($MA:NH_2$=1:0,5; maximum conversion=50%). The glass transition temperatures (Tg) of the reaction products are similar to the glass transition temperatures of poly(styrene-co-N-phenylmaleimide) (S/NPI) which was produced by radical polymerisation (see Figure 8).

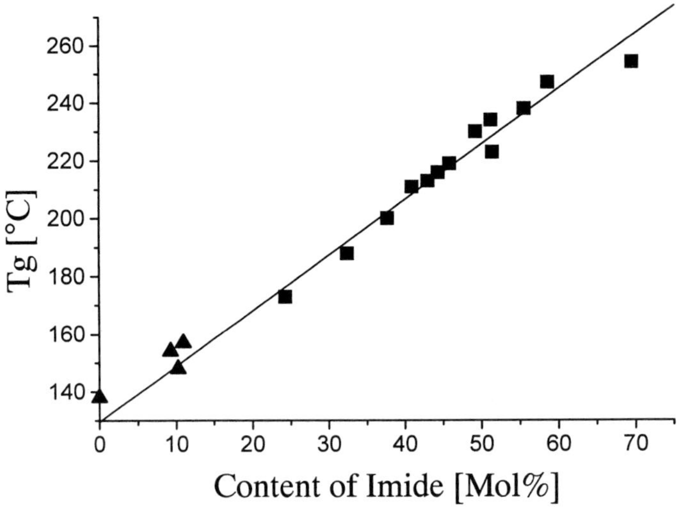


**Figure 8**: Tg-comparison of products from reactions with S/MA(84/16) and 3AP[▲] and products of S/NPI-copolymerisation[■]

Under the reported reaction conditions only the aminolysis can be observed when the additive 3AP is used. This is due to the low reactivity of the hydroxyl group. The use of 5AN enables to control aminolysis and alcoholysis. Its obvious that this reaction behaviour is due to the different nucleophile of the additives. 5AN reacts with both S/MA-16 and S/MA-32 in the expected way i.e. the alcoholysis is slower than the aminolysis. We found that the solubility is a good indicator for the degree of crosslinking. The decrease of the solubility is a function of the reaction time and the additive concentration (see Figure 9).

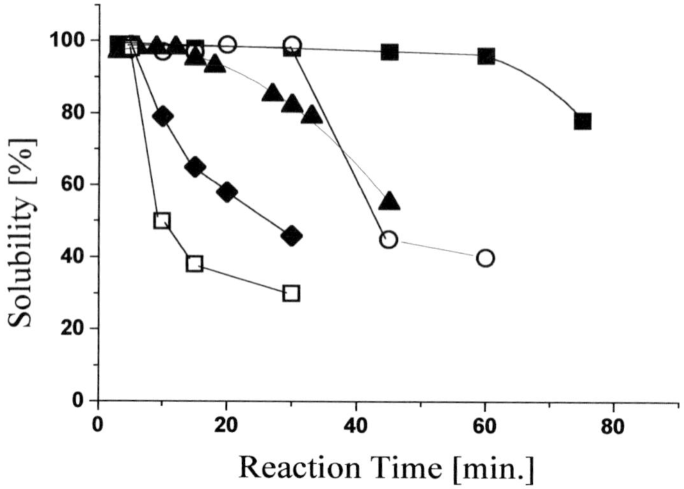


**Figure 9**: Solubility of reaction products from reactions with S/MA (84/16) and 5AN

[■]MA:$NH_2$=1:0,25; [O]MA:$NH_2$=1:0,5; [▲]MA:$NH_2$=1:1;

[◆]MA:$NH_2$=1:1,5; [□]MA:$NH_2$=1:2

## 6. Conversion of poly(styrene-co-acrylonitrile) (S/AN) with amino alcohols

The introduction of heterocyclic structures in polymers can be used as a basis for various further modification and treatment. Examples are reactive blending as well as the use of the new reactive centerpoints for draft polymerisation or as a basis for coatings or reactive paints.

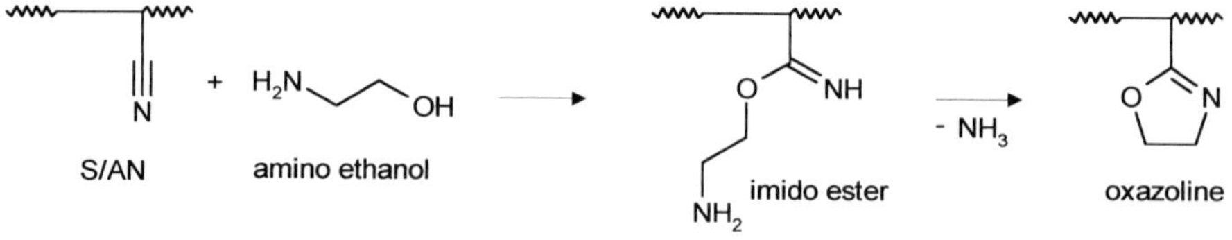


**Scheme 7**: Reaction of S/AN with amino ethanol leading to oxazoline products.

Seeliger and Witte [15,16] gave detailed descriptions of the reaction of nitriles and 2-amino alcohols to oxazolines (see Scheme 7). Polymer analogue reactions on the nitrile groups (AN) of S/AN in solution were realised from Hseigh et al. [17,18]. The transfer of this reaction into molten state eas done by Hu et al [19] and in our workgroup [2,3]. For aceotropic poly(styrene-co-acrylonitrile) we found that a reaction temperature of 190 °C is ideal. Even though Hu et al[19] favoured zinc acetate for the catalysis of the reaction, we found that dibutyltinn oxide gives higher yields. The used additives are arranged in Table 1. We put special interest in the influence of the following reaction parameters:

=> **additive concentration**

=> **catalyst concentration**

=> **reaction performance**

=> **structural influences of the additives.**

In Figure 10 the example of the additive 2-amino-2-ethylpropane-1,3diol (AEPD) shows that an increase of additive concentration results in an increase of conversion limit. The batch ratio of additive to catalyst was constant.

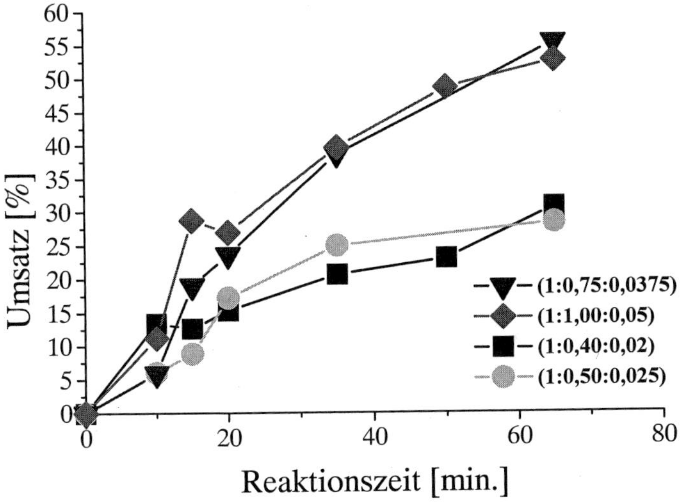


**Figure 10**: Reactions of S/AN with AEPD using various batch ratios [AN]:[additive]

As one of the most important reaction parameters we found the catalyst concentration.

The maximum conversation of reactions with S/AN and AEPD (AN:AEPD=1:1, catalyst concentration is varied) as a function of the catalyst concentration is shown in Figure 11. It is noteworthy that the coherence of maximum yield and catalyst concentration shows a maximum. This can be due to a deactivation of the nitrile group by excessing DBTO.

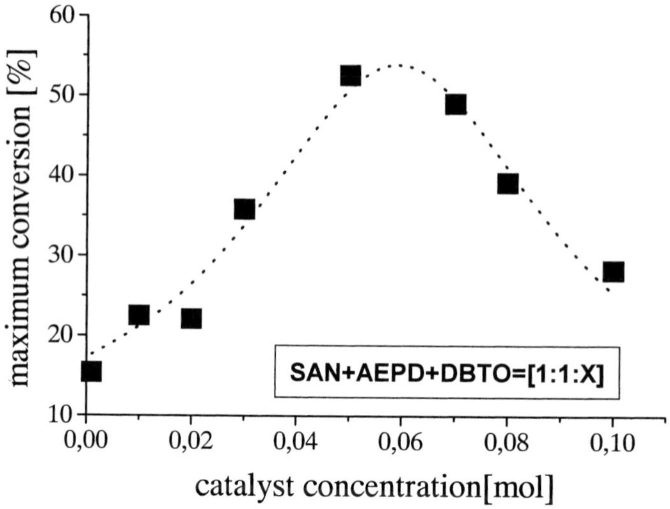


**Figure 11**: Coherence of maximum yields and catalyst concentrations of several DBTO-catalysed reactions of AN (S/AN) with AEPD.

Through an improved reaction performance, that means a stepwise addition of the additive, even additives with a low boiling point, like for example, 2-amino ethanol (AE) ($b_p$=170 °C) which shows a conversion limit of 17-31 % under standard conditions [2,19] can yield up to 50% [6] (see Figure 12). Therefore the reaction performance is changed from batch to semi-batch.

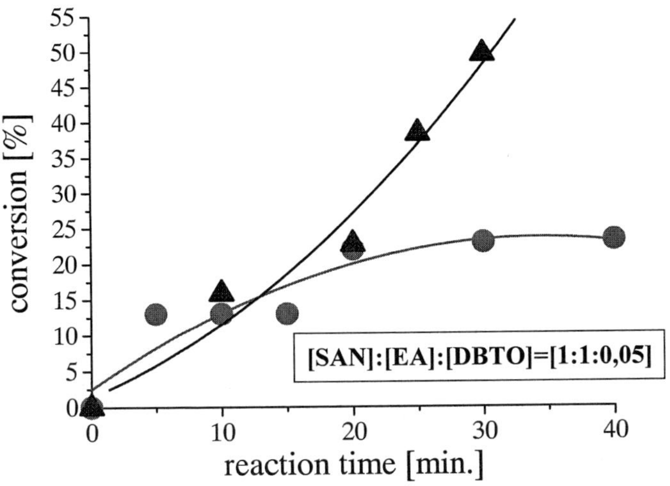


**Figure 12**: Influence of the addition mode of the additive on the conversion in DBTO-catalysed conversions of S/AN with AE (batch ratio: [AN:AE:DBTO]=[ 1:1:0,05]).

Figure 13 shows the structural influence of the additive. Reactions with S/AN and substituted 2-amino alcohols result in higher yields than with unsubstituted (AE). The highest yield was reached with 2-amino phenol (2AP) while the conversions with AEPD and 2-amino-1-phenyl-1-propanol (APP) showed a similar course. We note the reason for this in the aromatic stabilisation of the reaction product of AN and 2AP. Actually the product is not an oxazoline but an oxazole.

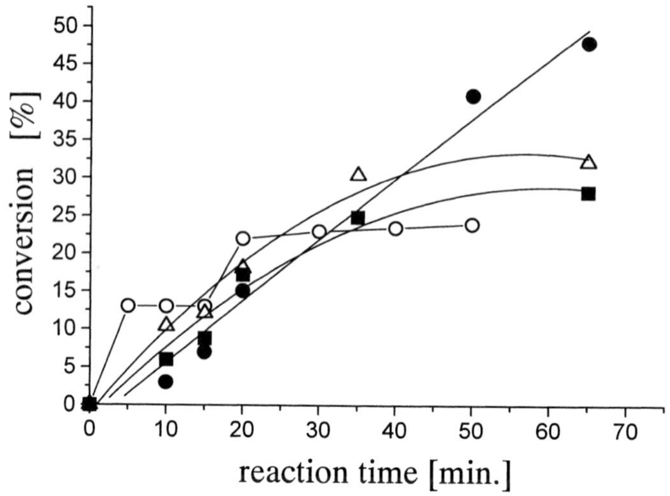


**Figure 13**: Coherence of conversion and reaction time in DBTO-catalysed reactions of AN (SAN) with AE,APP,AEPD and AP in the batch ratio of [AN : additive : DBTO] = [1:0,5:0,025] (○)AE/(Δ)APP/(■)AEPD/(●)2AP

The maximum yields of the reactions with the batch ratio of [AN:additive:DBTO]=[1:0,5:0,025] (Figure 13) show the following increasing line dependent to their product structure (Scheme 8):

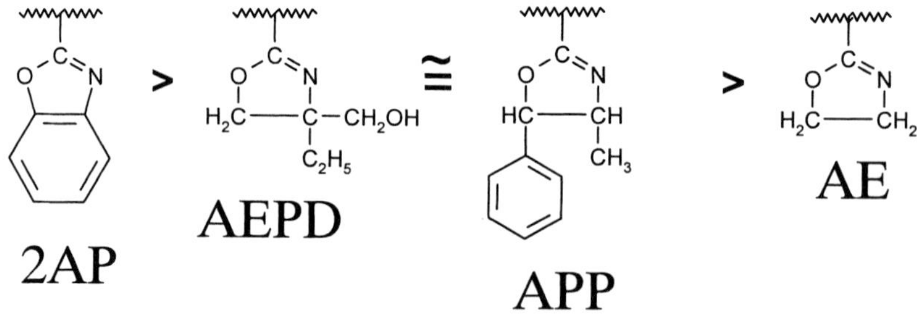


**Scheme 8**: Reaction products and their formation propabilities (experimental).

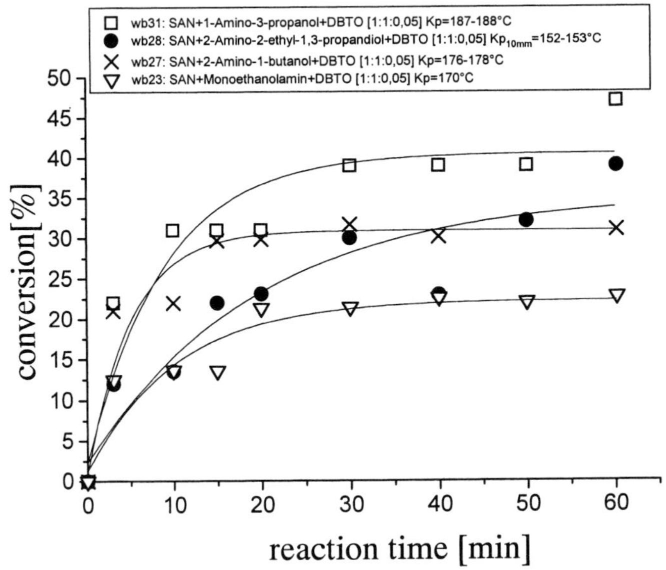


**Figure 14**: Conversion-time curve by variation of the additive [AN:additive:DBTO]=[1:1:0,05].

Using only aliphatic additives (Figure 14) results in the following decreasing line (Scheme 9):

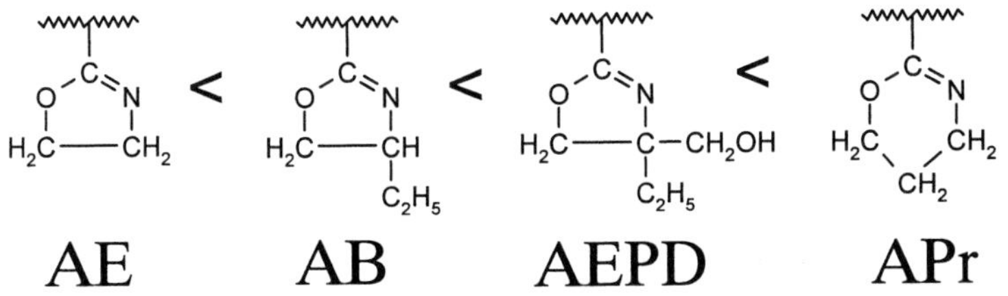


**Scheme 9**: Reaction products and their formation propabilities (experimental/aliphatic additives).

Therefore we cán state that even if the forming of substituted 5-ring-systems compared with unsubstituted is favoured, it is topped by the formation of 6-ring-systems (additive 3-amino propanol (APr)).

## Conclusion

Through the variation of the additives and catalysts during polymer modification the material properties of polymers can be changed in various manner. On the example of S/MA-modification, it can be shown that polymer modification has the tools to develop products with various new properties (for example with high thermal shape stability, polyelectrolytic and crosslinked products) out of simple industrial polymers. The introduction of heterocyclic structures in (S/AN) can be used as a basis for various further modifications and treatment leading to new polymer products with fascinating possibilities in use which have been only poorly investigated yet.

## References


1. N. Methfessel,G.Schmidt-Naake;Chem.Ing.Technik **70,**560(1998)
2. G.Schmidt-Naake,H.G.Becker,W.Becker;Angew.Makromol.Chem. **267**,63(1999)
3. H.G. Becker,G.Schmidt-Naake;Angew.Makromol.Chem. in press
4. G.Schmidt-Naake,M.Klak; Chem.Ing.Technik **71**,1418(1999)
5. G.Schmidt-Naake,M.Klak;Chem.Engineering and Technology (2000) in press
6. H.G.Becker,N.Merayo,G.Schmidt-Naake; Chem.Engineering and Technology (2000) in press
7. T,Otsu,S.Aoki,R.Nakatami;Makromol.Chem. **134**,331(1970)
8. I.Vermeesch,G.Groeninckx;J.Appl.Polym.Sci. **53**,1365(1994)
9. M.Lambla;Polym.Proc.Eng. **5**,297(1987)
10. Z.Song,W.E.Baker; J.Polym.Sci. Part **A30**,1589(1992)
11. M.Rätsch,V.Phien;Faserforsch.Textiltech. **26**,165(1975)
12. J.Pato,M.Azori,F.Tüdös; Makromol.Chem.Rapid Comm. **4**,25(1983)
13. G.Schmidt-Naake,M.Klak; Chem.Ing.Technik in preparation
14. R.C.P.Cubon; J.Polym.Sci. Part **C16**,367(1967)
15. W.Seeliger,H.Witte;Angew.Chem. **78**,913(1966)
16. H.Witte,W.Seeliger;Liebigs Ann.Chem. 996(1974)
17. D.T.Hseih,D.G.Pfeiffer;J.Appl.Polym.Sci. **56**,1667(1995)
18. D.T.Hseih,D.N.Schulz,D.G.Pfeiffer;J.Appl.Polym.Sci. **56**,1673(1995)
19. G.H.Hu,R.Scaffaro,F.La Mantia;J.Macromol.Sci.,Pure Appl.Chem. **A35**,457(1998)

*Macromol. Symp.* ***2001,*** *163, 235-250*


# Polymer Blends: Materials with versatile properties

M. Weber

BASF Aktiengesellschaft, Polymers Research Laboratory, 67056 Ludwigshafen, Germany

SUMMARY: During the last fifty years blending of dissimilar polymers has been a major path to tailor materials with new properties in industry. Especially in the area of engineering thermoplastic materials this approach has led to a significant number of large volume products, like Polyphenyleneether/High Impact Polystyrene-blends (PPE/HIPS), Polycarbonate/Styrenics-blends (PC/ABS), Polycarbonate/Polybutyleneterephthalate-blends (PC/PBT) and Polyamide/Polyphenyleneether-alloys (PA/PPE). The commercial success of these materials is mainly related to their unique combinations of properties, which enables their use in a multitude of applications.

## Introduction

During the last fifty years blending of dissimilar polymers has been a major path to tailor materials with new properties in industry[1-3].

There are several reasons for the broad acceptance of the blending technology as tool for the development of new materials. In the area of polymeric materials, usually a combination of properties is needed to fulfill the requirements of a particular application. For applications in different fields, like the automotive and electrical industry, not only mechanical properties like stiffness, strength and toughness are important, but also the thermal behavior (continuous use temperature, heat distortion temperature), processing properties (melt flow, melt stability), electrical properties, chemical resistance and flame resistance have to be considered.

If the development of a new material starts from new monomers or co-monomers, several properties can be optimized to achieve the anticipated performance. The most important properties are the chemical structure, molecular weight and molecular weight distribution, and the chain topology. For the development of new polymer blends, the already mentioned properties can be changed for each blend component, but additionally, the blend composition and in the case of immiscible polymers, also the morphology can be tuned to obtain the desired set of properties.

 CCC 1022-1360/00/$ 17.50+.50/0

Since the development of polymer blends usually starts from well known polymers, materials with interesting properties can be obtained with comparatively low investment for product development. If already existing product lines, usually twin screw extruders, are available, flexible production at a competitive cost level is already possible for smaller quantities (Tab. 1).

Table 1. Advantages of the blending technology for the product development

| **Development of new** | |
|---|---|
| **Polymers/Copolymers** | **Polymer Blends** |
| • Synthesis of new monomers, catalysts,...<br><br>• Development of a new polymerisation process<br><br>• TSCA-approval of monomer/polymer | • well-known polymers<br><br>• Development of a formulation |
| • construction of plants for monomer/polymer production | • melt mixing equipment |
| => high investment<br>=> long time to market | => low investment<br>=> short time to market |

The possibilities of the blending technology have already been recognized in the nineteenth century, when T. Hancock applied for a patent, disclosing mixtures of natural rubber and gutta-percha as materials for the improvement of textiles[4)]. The further development of polymer blends was of cause related to the availability of new polymers. The main developments in the field of polymer materials are summarized in table 2[2,5)].

Especially in the area of engineering thermoplastic materials this approach has led to a significant number of large volume products, like Polyphenyleneether/High Impact Polystyrene-blends (PPE/HIPS), Polycarbonate/Styrenics-blends (PC/ABS), Polycarbonate/ Polybutyleneterephthalate-blends (PC/PBT), and Polyamide/Polyphenyleneether-alloys

Table 2. History of polymer blends, important developments

| Year | Polymer Blend | Properties |
|---|---|---|
| 1846 | Natural Rubber/Guttapercha | stiffness/toughness |
| 1942 | PVC / NBR | NBR as plasticizer for PVC |
| 1960 | PPE / PS | melt processibility |
| 1969 | ABS / PVC | processibility, flame resistance |
| 1976 | PET / PBT | dimensional stability |
| 1977 | PC / ABS | impact strength |
| 1979 | PC / PBT | toughness, chemical resistance |
| 1982 | PA / Polyolefin | toughness, barrier properties |
| 1983 | PA 6,6 / PPE | high HDT |
| 1984 | PA / ABS | flow, toughness |
| 1987 | PBT / ASA | dimensional stability |
| 1990 | PEI / PC | flow |
| 1996 | PC / LCP | dimensional stability |
| 1998 | PES / PA | mechanical strength |

(PA/PPE). The commercial success of these materials is mainly related to their unique combinations of properties, which enables their use in a multitude of applications. The actual consumption of the most important polymer blends already reached the amount of 300.000 t/a (Tab. 3)[6,7].

Although some of the materials are on the market for more than 25 years, their sales volumes still increases with higher growth rates than the sales of the most engineering thermoplastic materials[6].

Table 3: Sales volumes and producers of important polymer blends

| Type | Components | Sales in `98 | Producer |
|---|---|---|---|
| miscible | PPE / HIPS | 305 000 | GEP, Asahi, BASF |
| | PSAN / PMMA / MBS | 25 000 | Toray, BASF |
| compatible | PC / Styrenics | 320 000 | GEP, Bayer, Dow |
| | PC / Polyester | 60 000 | GEP, Bayer, BASF |
| alloy | PPE / PA | 25 000 | GEP, Sumitomo |
| | ABS / PA | 20 000 | BASF, Bayer |

The importance of polymer blends for both industrial and academic laboratories is also reflected by the large number of patent applications and publications related to this topic (Fig. 1)[8].

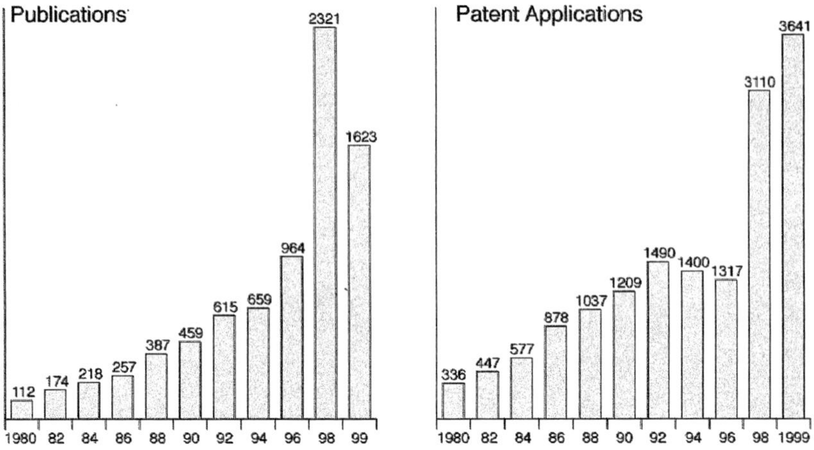


Fig. 1: Number of publications and patent applications related to polymer blends during the period from 1980 to 1999.

Especially during the last five years a tremendous increase in the number of publications as well as the number of patent applications could be observed.

## Classification of polymer blends

The multitude of existing polymer blends can be distinguished by their phase behavior. The thermodynamics of polymer-polymer-mixtures can be treated by a simple lattice model[9,10]. This treatment illustrates the small contribution of the mixing entropy to the free energy of mixing in polymer-polymer-mixtures. Therefore, the enthalpy of mixing is mainly responsible for the miscibility of polymer mixtures. The miscibility behavior of different polymers is subsequently related to interactions between the components, which could be of different nature, for example dipole-dipole-interactions, ionic-interactions or hydrogen bonding[11].

Table 4: Classification of polymer blends

| blend type | systems |
|---|---|
| miscible | PPE/PS, PVC/PBT |
| compatible | PC/SAN; PBT/SAN, PES/PC |
| immiscible | PS/PB; PA/SAN; PA/PPE; PA/PP; PS/PE |
| alloy | PPE/PA; PA/ABS; PC/PBT |

The whole number of polymer blends can therefore be divided in four groups (Table 4).

The first group that has to be mentioned are the miscible blends, where the chain segments of the different chains are miscible on a molecular level. Such kinds of blends have only a single glass transition temperature (Tg), which mainly depends on the composition (Fig. 2a). PPE/PS (Polyphenyleneether/Polystyrene)-blends are considered to be the most important example for the group of miscible blends.

Systems that are either partially miscible or completely immiscible, but offer attractive performance are often designated as compatible polymer blends. These blends usually have two glass transition temperatures, which may slightly deviate from the Tg´s of the blend components (Fig. 2b). The deviation of the glass transition temperatures from the Tg´s of the blend components might be different and depend on the partial miscibility of each component in the other. On a microscopic scale these polymer blends have a phase separated structure (morphology) which could be of different nature, depending on the composition of the blends. Usually the major component forms the matrix phase, wherein particles of the minor phase are dispersed. In the area of the 1/1-mixture, a bi-continuous morphology may exist.

The size of the dispersed particles is related to the interfacial tension and the viscosity ratio between the matrix and the dispersed phase[12)]. Usually the particle size of the dispersed phase is in the range of 1 to 5µm. The most popular system belonging to this group are PC/Styrenics-blends.

The largest group are the immiscible blends, having a completely phase separated structure. Therefore, the glass transition temperatures of the components in the blends are exactly the same as for the pure components (Fig. 2c). Immiscible polymer blends usually have a coarse morphology with particles of a size up to several microns.

The nature of the interface is a main issue for the mechanical performance of the polymer blends. The lack of interfacial strength in the immiscible blends leads to adhesive failure and poor mechanical properties.

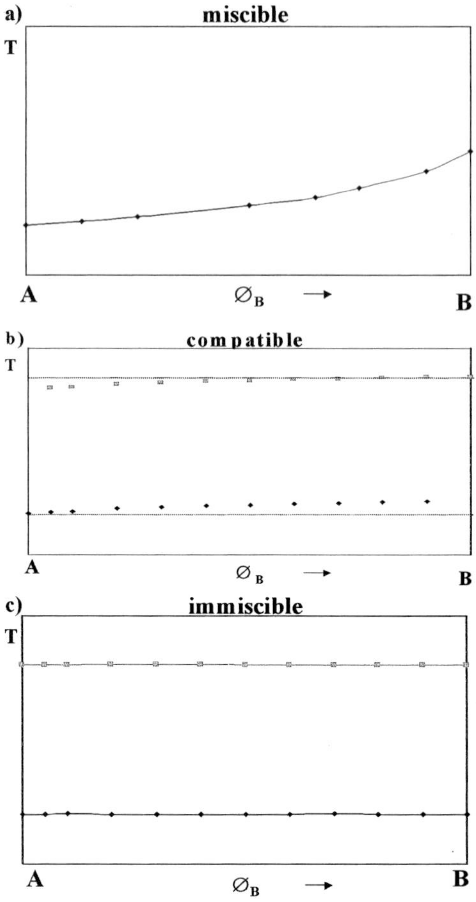


Fig. 2: Tg/composition-relationship for miscible, compatible and immiscible polymer blends

The morphology of immiscible blends can be modified by the addition of compatibilizers, which act like emulsifiers in oil/water mixtures. In polymer blends the compatibilizers are usually copolymers (block – or graft copolymers) consisting of different segments which are miscible with the respective components of the blend. During the melt mixing procedure the compatibilizer acts as polymeric surfactant, reducing the interfacial tension between the immiscible polymers, which results in a significant size reduction of the dispersed particles (Fig. 3). Since the surface of the dispersed particles is covered by the compatibilizer, the coalescence rate of the dispersed particles is tremendously reduced, keeping the morphology of the material stable during the later stage of mixing or subsequent processing steps[13-15].

In the bulk state, the compatibilizers provide a strengthening of the interface, which is an important issue for the toughness of multiphase materials[16-18].

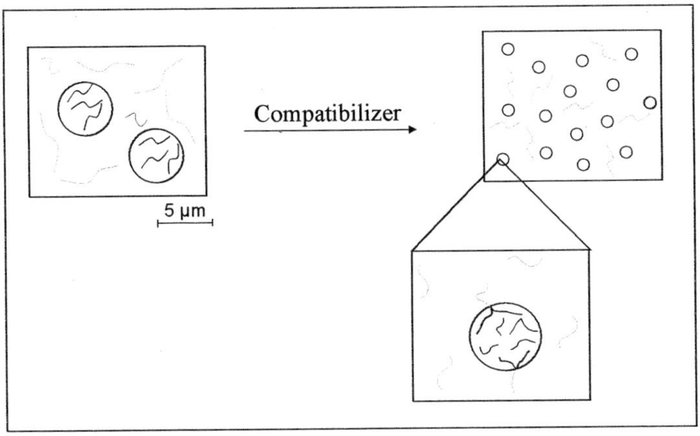


Fig. 3: Schematic representation of the compatibilizing effect of graft copolymers

## Important Engineering Polymer Blends and Alloys

## Polyphenyleneether/High Impact Polystyrene-Blends (PPE/HIPS-Blends)

Since 1960 it was known, that Polyphenyleneether and Polystyrene form homogeneous polymer mixtures[19]. By the addition of Polystyrene it was possible to improve the processibility of Polyphenyleneether and use these mixtures as thermoplastic materials[20,21]. Already in 1965 PPE/PS and PPE/HIPS-blends were commercialized by General Electric Plastics (Noryl®).

All PPE/PS-mixtures have a single glass transition temperature, which depends on the blend composition[22]. This behavior allows the adjustment of the heat distortion temperature of PPE/HIPS-blends by the simple variation of the blend composition. Therefore, one of the major disadvantages of Polystyrene, the low heat distortion temperature of approximately 90°C, can be significantly improved by the addition of PPE. Actually, some of the available PPE/PS-blends offer heat distortion temperatures of up to 190°C.

In order to improve the ductility and solvent resistance of PPE/PS-blends usually HIPS is used as blend component in commercial products. Therefore TEM-images of PPE/HIPS-blends mainly reveal the structure of the rubber particles (Fig. 4) of the HIPS-component. The further addition of Styrene-Butadiene-block copolymers (SB, SBS) or hydrogenated block copolymers (SEBS) is another tool to improve the toughness of these materials.

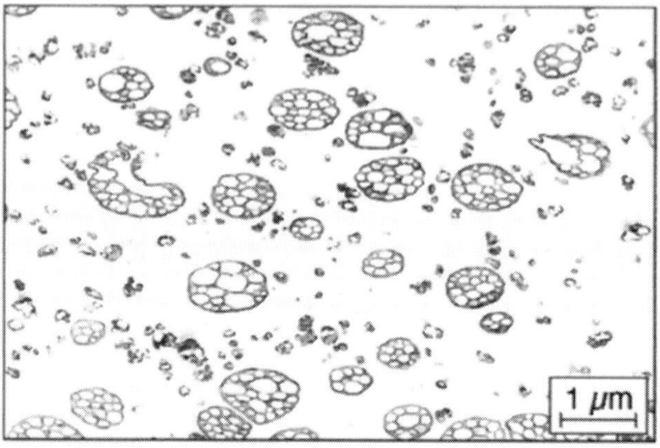


Fig. 4: TEM-image of a PPE/HIPS-blend stained with $OsO_4$

One of the major advantages of Polyphenyleneether is the low flammability. This property is maintained in the PPE/HIPS-blends to a large extent, therefore the addition of just 5 to 10 wt.-% of phosphate esters allows the production of materials with low flammability. These formulations fulfill the Underwriter Laboratories burning tests (UL 94) without the further addition of halogenated additives. This property is a prerequisite for the use of PPE/HIPS-blends in electrical and electronics applications.

As completely amorphous materials PPE/HIPS-blends also offer high dimensional stability, which allows the use up to temperatures near the glass transition temperature of the particular product without significant dimensional changes. Other advantages of PPE/HIPS-blends are the low moisture uptake and the good hydrolytical stability.

Due to this interesting combination of properties, PPE/HIPS-blends are excellent materials for a broad range of applications. The initial properties can be further diversified by the addition of fibers or fillers to enhance the stiffness and strength, or by the incorporation of other additives (anti-static agents etc.). Main fields of applications are any kind of housings in the electrical industry, automotive interior (dash board, radiator grills), machinery (pumps parts etc.).

## Polycarbonate/Styrene-Copolymer-Blends (PC/Styrenics-Blends)

Although Polycarbonate (PC) and Poly(styrene-co-acrylonitrile) (PSAN) are immiscible or only partially miscible, the mechanical properties of appropriate rubber toughened blends are excellent and exceed some of the properties of the blend components to a certain extend[5)].

PC/ABS-blends were first commercialized (Cycloy®) in 1967. Compared to Polycarbonate, PC/ABS-Blends offer improved processibility, chemical resistance and toughness. On the other hand, the heat distortion temperature as well as the toughness of the ABS is substantially improved by the addition of Polycarbonate[5)].

Different interpretations for the phase behavior of PC/PSAN-blends can be found in the literature. By DSC-measurements, usually a small reduction of the PC glass transition temperature and a small increase of the PSAN Tg can be detected for PC/PSAN-mixtures. While some authors count this behavior as an indication for partial miscibility[23,24)], other publications discuss about the influence of low molecular PSAN and PC for this result[25,26)].

The phase behavior is also influenced to a certain extend by the AN-content of the PSAN. The best phase adhesion between PSAN and PC is observed with PSAN having an AN-content of about 25 wt.-%[25)].

Due to the phase separated structure of PC/PSAN-blends there is no linear relationship between the heat distortion temperature and the composition of the materials. The heat distortion temperatures of the blends are determined by the nature of the matrix phase. If the PC-content exceeds 55 - 60 wt.-%, PC forms the matrix phase (Fig. 5, grey) wherein the ABS-particles are dispersed. In the dispersed ABS-phase the PSAN-matrix and the rubber particles (dark spherical particles) can be distinguished (Fig. 5). Most commercial products have this kind of morphology, which also gives rise to the high impact strength of these blends[5)].

As PPE/HIPS-blends, PC/ABS-blends also offer low flammability. In order to pass the UL-94 burning tests, Phosphate esters (Triphenylphosphate, Resorcinoldiphenyldiphosphate, etc.) are applied as flame retardants with loadings up to 15 wt.-%. Since these compounds are miscible with PC as well as ABS, the Tg´s of both phases are significantly suppressed, leading to a reduced heat distortion temperature. For certain applications a high heat distortion temperature (> 125°C) is necessary, therefore also products with halogenated flame retardants are available on the market. As halogenated flame-retardants brominated Polycarbonates or Polyhydroxyethers are used.

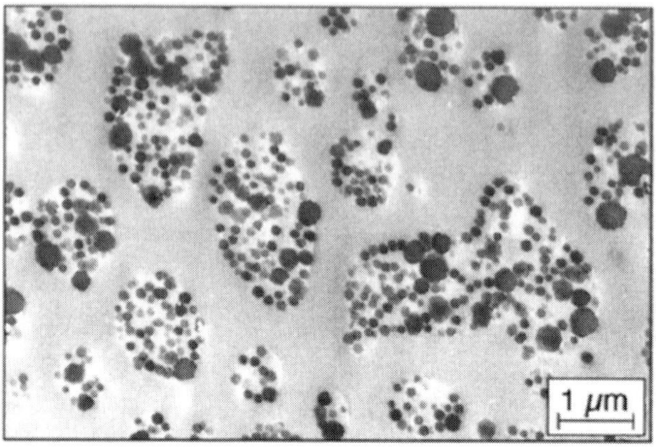


Fig. 5: TEM-image of a PC/ABS 60/40-blend stained with $RuO_4$

Due to the high ductility of PC/ABS-blends, these materials are widely used for automotive applications like automotive interior, mirror housings, panels, spoilers.

A major drawback of PC/ABS-blends is their low weatherability, which is of course due to the presence of Polybutadiene rubber. To improve this deficiency, other kinds of PC/Styrenics-blends were developed. In these blends, rubbers without olefinic double bonds (Acrylate-rubber: ASA, Ethylene-propylene-rubber: AES) are used as impact modifier. Especially the PC/ASA-blends have excellent weatherability and can be used for out-door applications.

## Polycarbonate/Polybutyleneterephthalate-Blends (PC/PBT/MBS-Blends)

Polycarbonate/Polybutyleneterephthalate-blends are a combination of an amorphous and a semi-crystalline material. The excellent toughness and dimensional stability of PC is combined with the good flow and the high solvent resistance of PBT. Both polymers are immiscible, but since PBT contains traces of titanium-catalysts from the polycondensation process, transesterification reactions between the two polymers lead to the formation of

Polyestercarbonates during the melt mixing process[27-29]. The Polyestercarbonates improve the compatibility between both polymers, but the formation of Polyestercarbonates has to be controlled by the addition of scavengers for the titanium-species, to avoid further reactions during compounding and processing, which finally would lead to almost amorphous materials with undesirable properties.

Therefore, the control of the transesterification process during compounding and processing of PC/PBT-materials is a critical issue. Usually Phosphites or Phosphates are used as transesterification inhibitors[30,31].

The influence of the transesterification inhibitors can be seen from DSC measurements and TEM-images (Fig. 6 and 7). In figure 6 the DSC-traces of a PC/PBT/MBS-blend with and without a transesterification inhibitor after three heating cycles from 30 to 290°C are depicted. It can be easily seen that the sample without transesterification inhibitor has a very broad melting peak for the PBT-phase compared to the sample with Tetrakis- (2,4-di-tert.butylphenyl) -4,4′-diphenylenediphosphonite as transesterification inhibitor.

In the TEM-image of the PBT/PC/MBS-blend without transesterification inhibitor (Fig. 7, right hand side) no distinct PC-phase is visible after the thermal treatment. This is further evidence for the formation of Polyestercarbonates.

As can be seen from figure 7, the MBS-rubber in PC/PBT/MBS-blends is exclusively located in the PC-phase. The MBS-rubber mainly boosts the low temperature impact performance of these blends.

Due to the excellent balance of properties like high toughness, good dimensional stability, high heat distortion temperature and good flow properties, PC/PBT/MBS-blends are used for large exterior parts in the automotive industry.

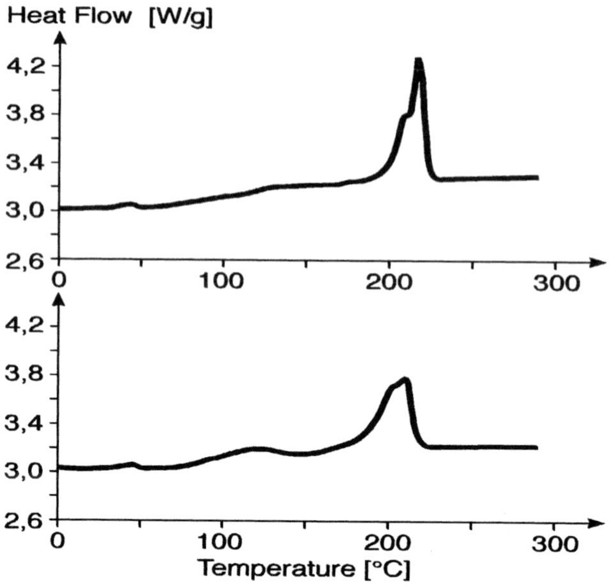


Fig. 6: DSC-traces of PC/PBT/MBS 48/42/10-blends after three heating cycles from 30°C to 290°C with a heating rate of 20 k/min; lower picture) without transesterification inhibitor, upper picture) with 0,5 wt.-% of Tetrakis-(2,4-di-tert.butylphenyl)-4,4´-diphenylenediphosphonite

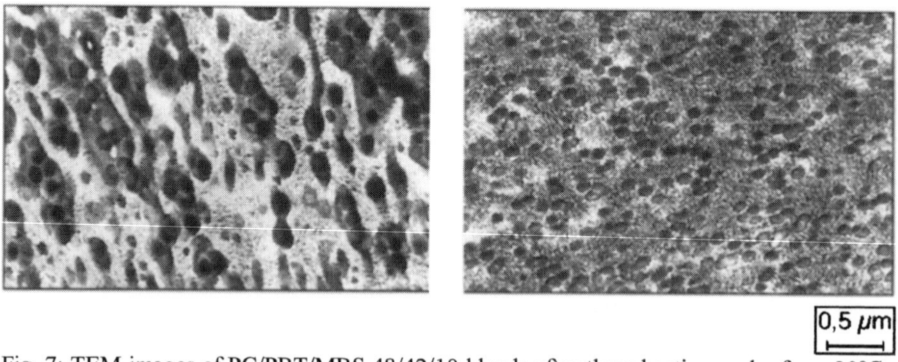


Fig. 7: TEM-images of PC/PBT/MBS 48/42/10-blends after three heating cycles from 30°C to 290°C with a heating rate of 20 k/min; right side) without transesterification inhibitor, left side) with 0,5 wt.-% of Tetrakis-(2,4-di-tert.butylphenyl)-4,4´-diphenylenediphosphonite

## Polyphenyleneether/Polyamide-Alloys

Another combination of amorphous and semi-crystalline polymers are the Polyphenyleneether/Polyamide-alloys (PPE/PA). As already discussed previously, PPE offers several interesting properties like high glass transition temperature, low moisture up-take, and high dimensional stability. On the other hand, this material has low chemical resistance and bad processibility. Therefore, a combination with a polymer like Polyamide 6,6, which has good chemical resistance, excellent melt processibility, and good toughness could offer interesting properties.

Blends of PPE and PA 6,6 are completely immiscible and have very unsatisfactory mechanical properties. The main issue in this system was to find an appropriate compatibilizer. Several different ways to compatibilize PPE and PA 6,6 are described in the literature[32-35)]. The most interesting way is the use of modified PPE, which is obtained by grafting of functionalized species like Maleic anhydride or Fumaric acid onto the PPE-chain in the melt. The modified PPE can subsequently react with PA 6,6 during the compounding procedure in order to form graft copolymers, composed of a PPE-backbone with PA 6,6-grafts. This process leads to a significant decrease of the particle size of the dispersed PPE-particles as can be seen from figure 8. Usually PA 6,6 (appears bright in TEM-image) forms the matrix phase whereas PPE forms the dispersed phase (grey particles). The average particle size of the dispersed phase is below 1 µm, which is a prerequisite to achieve an appropriate toughness level. In order to boost the toughness of the system, these alloys furthermore contain hydrogenated Styrene-Butadiene-Styrene-block copolymers. Due to the presence of the Styrene-blocks, these impact modifiers are located exclusively in the dispersed PPE-particles (Fig. 8).

The first PPE/PA-alloy on the marked was the so-called Noryl®GTX. This product was introduced to the market in 1986.

Due to the high Tg of the PPE-phase and the semi-crystalline nature of polyamide 6,6, these alloys have a high heat distortion temperature of up to 190°C, which allows the direct painting of parts molded from this alloy at temperatures up to 170°C in usual paint lines used for steel parts (on-line painting). Therefore PA/PPE-alloys are widely used as material for large exterior parts in the automotive industry[5,36)].

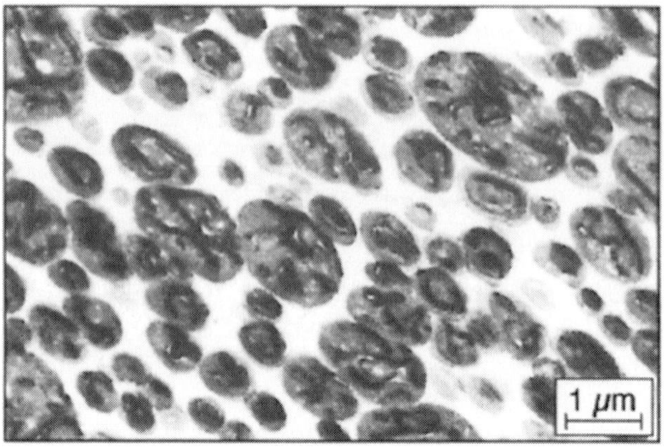


Fig. 8: TEM-image of a PA 6,6/PPE/SEBS-alloy stained with $RuO_4$

## New Developments

During the last years several new polymer blends appeared on the market. As examples PC/LCP-, PES/PA- and PEI/PET-blends will be discussed briefly.

The combination of Polycarbonate and a thermotropic liquid crystalline Polyester (LCP) yields materials with excellent dimensional stability and good flow. Due to these properties, PC/LCP-blends can be used as materials for connectors, where requirements regarding to dimensional accuracy, heat distortion, and melt flow have to be fulfilled[37].

Fiber reinforced blends of Polyethersulfone (PES) and Polyamide offer excellent mechanical strength and high melt flow. The good heat resistance of PES is still maintained in these products. Compared to PES, PES/PA-blends offer a significantly improved performance/price-ratio. Fiber reinforced PES/PA-blends are appropriate for under the hood-applications in the automotive industry[38].

Mixtures of the high performance polymer Polyetherimide (PEI) and Polyethylene-terephthalate are marketed under the tradename Ultem ATX. Compared to pure PEI, this material offers improved chemical resistance, higher melt flow and of course a significant cost reduction[39].

## Conclusions

Due to the tremendous advantages regarding the investment for product development and production, the blending technology is still used extensively for the development of new materials both in industrial and academic laboratories. In the area of engineering thermoplastic materials several products (PPE/HIPS, PC/ABS, PC/PBT/MBS, PA/PPE) have already reached large sales volumes, but most of these products still grow faster than the usual engineering plastics. This is of cause due to their unique combinations of properties and the large number of possibilities to fine tune the overall performance.

## Acknowledgements

The contributions of Dr. W. Heckmann and Dr. I. Henning are acknowledged. The author also wishes to thank B. Cunningham and H. Barosch for their technical assistance.

## References

1. D.R. Paul, S. Newman, *Polymer Blends,* Academic Press, New York 1978
2. L.A. Utracki, *Polymer Alloys and Blends*, Hanser, München, Wien, New York 1989
3. L.A. Utracki, *Encyclopaedic Dictionary Of Commercial Polymer Blends* ChemTec Publishing, Toronto 1994
4. Engl. Pat. 11,147, (1846) T. Hancock
5. L. Bottenbruch, *Technische Polymer-Blends,* Kunststoff-Handbuch 3/2, G.W. Becker, D. Braun (Eds.), Hanser, München, Wien 1993
6. H. Warth, D. Wittmann, *Kunststoffe* 10/**89**, 126 (1999)
7. J. Braat, J. Liska, *Kunststoffe* 10/**89**, 118 (1999)
8. Literature survey with CAS, M. Wijnands, BASF, March 2000
9. M.L. Huggins, *J. Chem. Phys.* **9**, 440 (1941)
10. P.J. Flory, *J. Chem. Phys.* **9**, 660 (1941)
11. O. Olabisi, L.M. Robeson, M.T. Shaw, *Polymer-Polymer-Miscibility* Academic Press, New York 1979
12. S. Wu, *Polym. Eng. Sci.* **27**, 335 (1987)
13. C.E. Scott, C.W. Macosko, *J.Polym. Sci., Part B* **32**, 205 (1992)
14. C.E. Scott, C.W. Macosko, *Polymer* **35**, 5422 (1994)

15. A. Nakayama, T. Inoue, A. Hirao, P. Guegan, P. Khandpur, C.W. Macosko, *Polym. Prepr.* **34**(2), 840 (1993)
16. H.R. Brown, *Macromolecules* **24**, 2752 (1991)
17. C. Creton, E.J. Kramer, G. Hadziioannou, *Macromolecules* **24**, 1846 (1991)
18. J. Washiyama, C. Creton, E.J. Kramer, *Macromolecules* **26**, 2928 (1993)
19. US 3,063,851 (1962), General Electric, E.M. Boldebuck
20. F. E. Karasz, W.J. MacKnight, J. Stoelting, *Polym. Prepr. Chem. Soc. Div. Polym. Chem.* **14**, 357 (1970)
21. W.J. MacKnight, J. Stoelting, F.E. Karasz, *Adv. Chem. Ser.* **99**, 29 (1971)
22. W.M. Prest, R.S. Porter, *J. Polym. Sci. Polym. Phys. Ed.* **10**,1639 (1972)
23. J.D. Keitz, J.W. Barlow, D.R. Paul, *J. Appl. Polym. Sci.* **29**, 3131 (1984)
24. R.A Mendelson, *J. Polym. Sci., Polym. Phys. Ed.* **23,** 1975 (1985)
25. T.A. Callaghan, K. Takakuwa, D.R. Paul, A.R. Padwa, *Polymer* **34**, 3796 (1993)
26. M.J. Guest, J.H. Daly, *Eur. Polym. J.* **25**, 985 (1989)
27. J. Devaux, P. Godard, J.P. Mercier, *J.Polym.Sci., Polym. Phys. Ed.* **20**, 1875 (1982)
28. J. Devaux, P. Godard, J.P. Mercier, *J.Polym.Sci., Polym. Phys. Ed.* **20**, 1895 (1982)
29. A.N. Wilkinson, S.B. Tattum, A.J. Ryan, *Polymer* **38**, 1923 (1997)
30. A. Golovoy, M.-F. Cheung, K.R. Carduner, M.J. Rokosz, *Polym. Eng. Sci.* **29**, 1226 (1989)
31. EP 683 200 (1995) General Electric, R.C. Crosby, L.I. Flowers, R.R. Odle, J.L. De Rudder, Y.-G. Lin
32. US 4,315,086 (1982), Sumitomo Chemical Company Ltd., K. Ueno, T. Maruyama
33. US 4,338,412 (1982), Sumitomo Chemical Company Ltd., K. Ueno, T. Maruyama
34. EP 129 825 (1985), General Electric, V. Abolins, J.E. Betts, F.F. Holub
35. US 5,357,003 (1994), General Electric, H.J.E. Smits, R. van der Meer, A.H.L. Groothius
36. C. Verpy, J.L. Gacougnolle, A. Dragon, A. Vanlerberghe, A. Chesneau, F. Cozette, *Progr. Organic Coatings* **24**, 115 (1994)
37. M. Roemer, A. Neupauer, *Kunststoffe* **86**, 1310 (1996)
38. M. Weber, U. Eichenauer, J. Queisser, *Kunststoffe* **88**, 1472 (1998)
39. M.Huff, *Polymer News* **22**, 290 (1997)